Bioarchaeology and Climate Change

Bioarchaeological Interpretations of the Human Past:
Local, Regional, and Global Perspectives

UNIVERSITY PRESS OF FLORIDA

Florida A&M University, Tallahassee
Florida Atlantic University, Boca Raton
Florida Gulf Coast University, Ft. Myers
Florida International University, Miami
Florida State University, Tallahassee
New College of Florida, Sarasota
University of Central Florida, Orlando
University of Florida, Gainesville
University of North Florida, Jacksonville
University of South Florida, Tampa
University of West Florida, Pensacola

Bioarchaeology and Climate Change

A View from South Asian Prehistory

GWEN ROBBINS SCHUG

Foreword by Clark Spencer Larsen

University Press of Florida

Gainesville · Tallahassee · Tampa · Boca Raton
Pensacola · Orlando · Miami · Jacksonville · Ft. Myers · Sarasota

22 21 20 19 18 17   6 5 4 3 2 1

First cloth printing, 2011
First paperback printing, 2017

Library of Congress Cataloging-in-Publication Data
Robbins Schug, Gwen.
Bioarchaeology and climate change : a view from South Asian prehistory / Gwen
Robbins Schug ; foreword by Clark Spencer Larsen.
p. cm.
Includes bibliographical references and index.
ISBN 978-0-8130-3667-0 (cloth: alk. paper)
ISBN 978-0-8130-5412-4 (pbk.)
1. Daimabad Site (India) 2. Inamgaon Site (India) 3. Nevasa Site (India)
4. Copper age—India—Deccan. 5. Excavations (Archaeology)—India—Deccan.
6. Human remains (Archaeology)—India—Deccan. 7. Paleoclimatology—India—
Deccan. 8. Climatic changes—India—Deccan. 9. Deccan (India)—Antiquities.
I. Title.
GN778.32.I5S38 2011
954.'8—dc22   2011012174

The University Press of Florida is the scholarly publishing agency for the State
University System of Florida, comprising Florida A&M University, Florida Atlantic
University, Florida Gulf Coast University, Florida International University, Florida
State University, New College of Florida, University of Central Florida, University
of Florida, University of North Florida, University of South Florida, and University
of West Florida.

University Press of Florida
15 Northwest 15th Street
Gainesville, FL 32611-2079
http://www.upf.com

Contents

List of Figures   vii

List of Tables   ix

Foreword   xi

Preface   xiii

Acknowledgments   xvii

1. Origins   1

2. The Western Deccan Plateau: Environment and Climate   25

3. Archaeology at Nevasa, Daimabad, and Inamgaon   38

4. Demography   61

5. Estimating Body Mass in the Subadult Skeleton   82

6. Reconstructing Health at Nevasa, Daimabad, and Inamgaon   89

7. Conclusion   114

    Appendix A. Burials from Daimabad: Archaeological Context
    and Grave Goods   125

    Appendix B. Age Estimates for Subadults in Chalcolithic
    Samples   128

    Appendix C. Long Bone Lengths (mm) and Stature (cm)
    for Individuals with Dental Age Estimates (months)   143

    Appendix D. Midshaft Femur Cross-Section Measurements
    for All Individuals with Intact Femur Midshafts   147

    Appendix E. Stature and Body Mass Estimates
    for Deccan Chalcolithic Specimens   149

    Notes   151

    Bibliography   155

    Index   177

Contents

1. Introduction

2.

3.

4.

5.

6.

7. Conclusion

Figures

1.1. Map of archaeological sites in the Indus Age and the Deccan Chalcolithic 2

1.2. Map of Chalcolithic sites in Maharashtra during the second millennium B.C. 3

3.1. Aerial view of Nevasa 40

3.2. Aerial view of Daimabad 45

3.3. Aerial view of Inamgaon 49

4.1. Observed versus predicted GRR using estimates of GRR from my formula and Bocquet-Appel's hazard ratio 65

4.2. Observed versus predicted GRR for 98 model life tables 67

4.3. Cumulative frequency of perinatal age distribution using prior probabilities 73

4.4. Average number of offspring per woman (TFR) versus life expectancy at birth (e_o) in Deccan Chalcolithic samples 77

6.1. Scaling relationship between body mass and stature squared in the Denver sample 93

6.2. %CA at the femur midshaft for age in the Denver sample 95

6.3. Box plots for stature in the Denver and Deccan Chalcolithic samples 105

6.4. Box plots for body mass in the Denver and Deccan Chalcolithic samples 105

6.5. Box plots for body mass scaled to stature squared in the Denver and Deccan Chalcolithic samples 106

6.6. Scatter plot of body mass scaled to stature squared 108

6.7. Body mass index plotted against age 108

7.1. Summary of models to explain culture change in the Jorwe phase of the Deccan Chalcolithic 115

Tables

1.1. Results of prior research on growth profiles at Inamgaon
    (% individuals) 14

2.1. Paleoclimatic evidence for the Late Holocene in South Asia 30

3.1. Total number of seeds recovered and proportional representation of
    species at Inamgaon by layer 51

3.2. The faunal assemblage from Inamgaon: Minimum number of
    individuals (%) 52

4.1. Sample characteristics and estimates for GRR based on the
    proportion of infants (0–12 months) in the subadult (< 20 years)
    sample from 98 Female Model West life tables 66

4.2. Test of Robbins formula on 11 prehistoric skeletal samples
    in comparison with estimates for GRR using Bocquet-Appel's
    ratio 68

4.3. Age estimates from perinatal long-bone lengths
    (in lunar weeks) 71

4.4. Proportion of perinatal individuals in each gestational age
    category 72

4.5. Subadult versus adult deaths 74

4.6. Subadult deaths by age category 75

4.7. Estimates for GRR, crude birth rate, and life expectancy at birth
    for Deccan Chalcolithic samples 76

5.1. Descriptive statistics for the Denver Longitudinal Study
    sample 87

5.2. Equations for predicting body mass (kg) from femur torsional
    rigidity (J) (untransformed data) 87

6.1. Body mass index in contemporary and archaeological samples from
    Denver, Grasshopper Pueblo, and Kulubnarti 94

6.2. Demographic and LHPC profiles for Nevasa, Daimabad, and
    Inamgaon 99
6.3. Sample size and proportion of individuals with dental age
    estimates, complete femur lengths, and intact femur midshaft
    compact bone 102
6.4. Mean estimated stature 104
6.5. Mean estimated body mass 104
6.6. Mean estimated body mass index 104
6.7. Mann Whitney test for significant differences among the Deccan
    Chalcolithic samples (pooled sample of age categories 0–5) 107
6.8. Mann Whitney tests for significant differences among samples
    based on differences in subsistence (infants age 0–1.49) 109

Foreword

Climate change has long been invoked to explain extinction events, the rise and fall of civilizations, and alterations in productivity in the past. Bioarchaeology, the study of human remains from archaeological settings, offers an important opportunity to address key issues relating to climate and a range of other circumstances. In the last two decades, bioarchaeologists have investigated the consequences of key adaptive shifts on health and well-being in the Holocene. The global development of the foraging-to-farming transition has figured most prominently in these discussions, revealing a general decline in health and quality of life in virtually every location studied, including in South and East Asia.

In this book, Gwen Robbins Schug offers an important study addressing a fundamental question: How did climate change impact subsistence and health of human populations during the latter half of the second millennium B.C. in peninsular India? Toward the end of the Deccan Chalcolithic period in west-central India, farmers virtually abandoned agriculture as the primary means of food production. Many authorities have long thought that this reversal—back to foraging—was somehow linked to climate change, largely involving a presumed association between increased aridity and reduced agricultural productivity. Many authorities have also suggested that this subsistence transition could have led to an improvement in health status for the people of Inamgaon—because of the association between agricultural subsistence and poor health.

Robbins Schug's study confirms that the links between climate, new dietary adaptations, and biocultural outcomes are complex and must be approached from numerous avenues of investigation. Specifically, she uses new methods and new approaches to assess demography, growth, and development and then to infer nutritional status and health. Her research reveals that human populations adjusted and adapted at a highly local level to increasing levels of aridity, which began in the phase known as the Early Jorwe (1400–1000 B.C.). Rather than representing an environmental disaster, climate change was initially met with an increase in adaptive diversity that was a successful strategy for a time. However, by the end of the second millennium B.C., after 400 years of settlement growth, these villages were largely abandoned. At that point, there was a transition away from agricultural production and a deemphasis on drought-resistant barley as a staple crop. The people of Inamgaon began to rely more heavily on wild foods, sheep and goat pastoralism, and saline-tolerant crops. Schug's demographic and skeletal growth profiles indicate that quality of life and human health suffered from the collapse of a major portion of the subsistence base. These community-level effects are demonstrated in a compelling manner.

Robbins Schug's biocultural synthesis provides us with a new way of looking at the adaptive, social, and cultural transformations that took place in the Deccan Chalcolithic. The links between climate, subsistence, and biological change are not so straightforward as once thought. As Schug so wonderfully demonstrates, the climatic and bioarchaeological record in South Asia demonstrates highly dynamic circumstances and complex links between environment, culture, and biology.

Climate change is an important element for understanding human settlement and population change. However, this study makes clear that old models relying on simple links between climate and resource productivity are not sufficient for explaining the broad sweep of human adaptation. New models drawing on multiple elements of complexity are where new breakthroughs are being made in the growing picture of adaptation and change in the archaeological past.

Clark Spencer Larsen
Series Editor

Preface

With some exceptions for East Asian agricultural populations (Pi-etrusewsky and Douglas 2001; Oxenham, Thuy, and Cuong 2005; Oxen-ham and Tayles 2006; Domett and Tayles 2007), research has repeatedly demonstrated that reliance on agriculture in prehistory generally had negative impacts on human populations (Cohen 1977, 1984; Larsen 1995; Lambert 2000; Steckel and Rose 2002; Cohen and Crane-Kramer 2007). Agricultural populations in general suffer from an unhealthy reduction in dietary diversity as they begin to rely on one or two staple crops for the majority of their energy intake. This tendency to specialize also can lead to more frequent food shortages when crops fail due to climate, pes-tilence, or disease. In populations that emphasize agricultural activities for subsistence, we generally find high pressure demographic situations and high morbidity rates. Particularly in New World prehistory, large-scale dependence on maize agriculture has become synonymous with in-creasing population density at higher reproductive cost for women (high fertility and high infant mortality), nutritional deficiencies, increases in developmental stress episodes, declines in health status, increased con-sumption of cariogenic and phytate-laden cereals that promote dental diseases, increased prevalence of socio-sanitation problems, increased parasite load, more communicable and degenerative diseases, and some-times increased rates of interpersonal violence.

Because agriculture often comes at a cost for human populations, the widespread adoption of agriculture as a principle means of subsistence is a conundrum and its complexities warrant serious study. Contrary to popular perceptions, the transition to agriculture was not a simple "Neolithic Revolution" in which human populations began an inexorable march toward agricultural production 10,000 years ago, leaving behind the "solitary, poore, nasty, brutish, and short" lifestyles of their hunting and gathering ancestors (Hobbes and Macpherson 1968). Humans have a complex relationship with their environment and there are multiple pathways that have led to domestication of plants and animals (Harris and Childe 1992; Smith and Winterhalder 1992; Kennett and Winterhalder 2006). It is a process that has occurred many times in human prehistory, but it is not inevitable, progressive, or better than other lifeways. There is variation in how much and how fast human communities commit to food production; many societies have retained a mixed economy for hundreds of years after adopting agriculture. Others have adopted agricultural production as a principle mode of subsistence and then reversed course and abandoned their fields in favor of a lifestyle better suited to uncertainty or changing circumstances.

An example of the latter situation is found in peninsular India during the second millennium B.C. This book seeks to understand what changed between 1400 and 700 B.C. that led to the abandonment of agriculture and eventually to the abandonment of settled life in this region for hundreds of years. I will examine hypotheses about climate and culture change in prehistoric India with the goal of understanding how these changes affected health in the human population. For centuries, Deccan Chalcolithic people farmed drought-resistant barley and wheat. They raised cattle, sheep, and goats. They maintained hunting and foraging traditions and utilized the resources gathered from local lakes and forest habitats for subsistence, construction, and fuel. About 1000 B.C., the majority of these villages were deserted. Inamgaon persisted for 300 years.

Clearly lifestyles changed at Inamgaon after agriculture was abandoned. This book seeks to characterize this transition and to evaluate competing hypotheses about climate, culture, and subsistence changes during the Deccan Chalcolithic period using recent paleoclimate reconstructions, the archaeological record, and the human skeletal material as my sources of evidence. Through a comparison of demographic parameters and evidence for growth disruption in infants and children at three sites (Inamgaon, Nevasa, and Daimabad), I will infer the effects of environmental change on Chalcolithic people and characterize the circumstances in which villages were abandoned during the time known as the Early Jorwe phase. I will also seek to understand how it was that the people of Inamgaon persisted longer, into the Late Jorwe phase. And finally, I hope to understand something about the conditions in which they too abandoned their village.

Questions about human health and interactions with the environment three or four thousand years ago in India are interesting from an academic standpoint, but the insights we gain into the past can also be relevant in a contemporary context as we face the consequences of continued population growth, unsustainable lifestyles, degradation of local environments, and large-scale climate changes. It is particularly interesting to study bio-archaeology, climate, and culture change in India given that hundreds of millions of Indian people are currently living in rural villages. In many cases, these communities demonstrate remarkable continuity with the material culture of villages of 3,000 years ago. Human populations occupying semiarid regions at low latitudes will arguably face the greatest magnitude of effect from the current global climate changes. Understanding something about the successes and failures of the strategies employed by past peoples provides some historical context for the choices facing us today.

Climate change is a prime mover in evolution. Humans have a complex relationship to environment and the capacity to respond creatively to environmental challenges. Thus we are not constrained by climate, landscape, habitat, and carrying capacity in the same way as other species, but there can be no doubt that human populations are profoundly affected by environmental changes at the local and global level. In fact, there have been many examples throughout human history of civilizations collapsing due to inadequate responses to changing environmental circumstances (Diamond 2004). Having a longer view on the challenges, strategies, and consequences of human responses to the environment may prove helpful as we all develop strategies for dealing with contemporary and future interactions between human society and the environment. This book begins the inquiry into the strategies that have come before, the successes and failures of our predecessors.

Acknowledgments

I wish to express my sincere appreciation to Clark Larsen for inviting me to participate in this series and for advice concerning the manuscript. I particularly want to thank John Lukacs, who has remained a principle contributor to my enthusiasm and love for biological anthropology and for South Asian bioarchaeology. He has been a tireless supporter and friendly critic throughout this process. Very special thanks are also due to Subhash Walimbe, Veena Mushrif-Tripathy, Virendra Misra, Ravi Mohanty, Vasant Shinde, and Ismail Kellellu. Their familiarity with the issues, controversies, and details of Indian prehistory were of utmost importance in the conception of ideas presented here, and they have contributed immeasurably to the experiences that I have had doing fieldwork in India over the years. I have received so much advice and assistance from so many scholars over the years, but I would particularly like to thank Mark Cohen, Michael Pietrusewsky, Guy Tasa, Stephen Frost, J. Josh Snodgrass, Frances White, Louis Osternig, Chris Ruff, and Jeanne Pierre Bouquet-Appel for their advice on aspects of this research and the manuscript.

My immense gratitude is extended to Deccan College Post-Graduate Research Institute for providing materials for study and housing in the residence halls. This research was supported financially the American Institute of Indian Studies, George Franklin Dales Foundation, IIE Fulbright, and the University of Oregon. I would like to express gratitude specifically to all of the staff at AIIS and Fulbright in India for their support. I wish to acknowledge my department and colleagues at Appalachian State University who have given friendship, support, and assistance to me in accomplishing many goals. You all make my journey easier and more pleasant. Finally, I want to extend a huge thank you to my children and to my husband, Malcolm Schug—I love you.

1

◇◇◇◇◇◇◇◇◇◇◇◇◇◇◇

Origins

> The people to whom these ruined sites belonged, lacking posts, these many
> settlements, widely distributed, they, O Agni, having been expelled by thee, have
> migrated to another land.
>
> Taittiriya Brahmana (2.4.6.8)

The most widely known period in South Asian prehistory is the "Indus Age"—a term that encompasses settled life in Pakistan and north-western India from incipient agricultural production after 7000 B.C. to the beginning of the Iron Age about 1000 B.C. (Possehl 2002). The term "Indus civilization" refers to a time during the latter half of the third millennium (2400–1900 B.C.) when Indus sites went through a "mature" phase, best represented by the archaeological record at large urban centers such as Harappa, Mohenjo Daro, Lothal, and Kalibangan (see the Harappa web site at http://www.harappa.com/ for more information). Trade goods, technology, seals, symbols, systems, and ideas were shared by more than a thousand settlements along the banks of the Indus River system—from the seven rivers that made up its headwaters to its fertile deltas on the Arabian Sea (Fig. 1.1).

The mature phase of the Indus civilization ended circa 1900 B.C., and the succeeding post-urban phase lasted for most of the second millennium (1900–1000 B.C.) in northwest India and Pakistan. The post-urban phase was characterized by significant decentralization of the core Indus territory. Many of the large population centers were abandoned and an increasing number of small villages and encampments proliferated at the margins of Indus territory and beyond, to the western end of the Ganges-Jumna doab, southern Gujarat, and south into peninsular India. In the Ganges-Jumna region alone, the number of settlements increased from 218 during the mature Harappan to 853 in the post-urban phase (Possehl

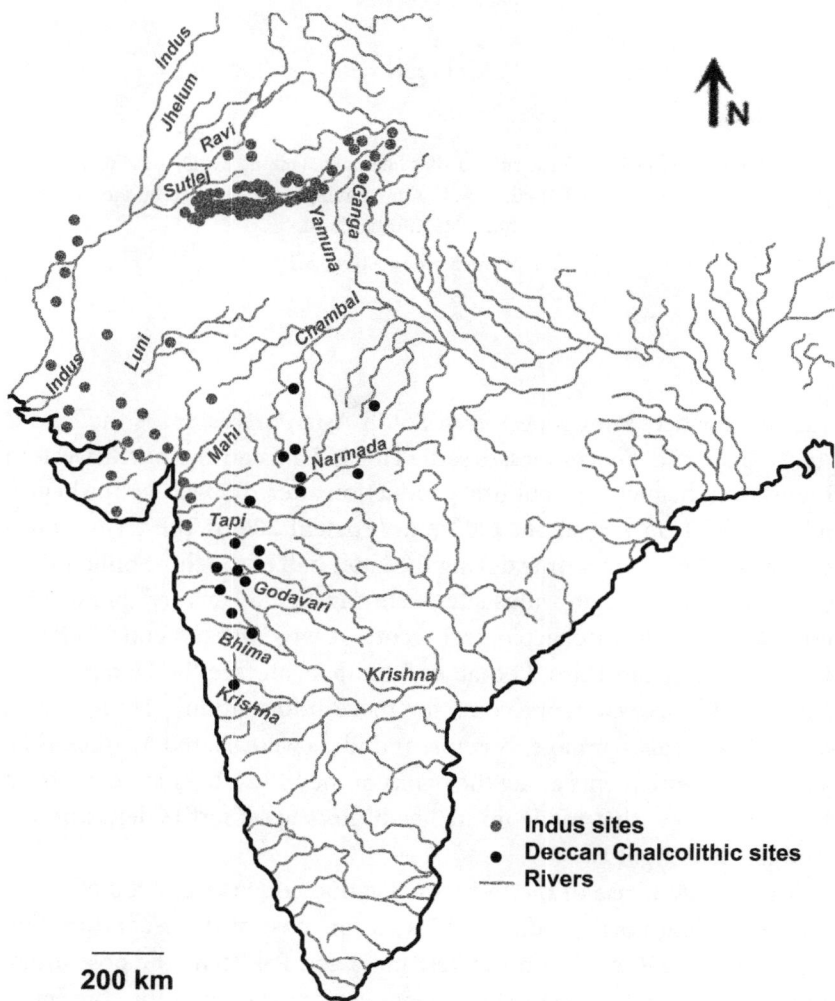

Figure 1.1. Map of archaeological sites in the Indus Age and the Deccan Chalcolithic.

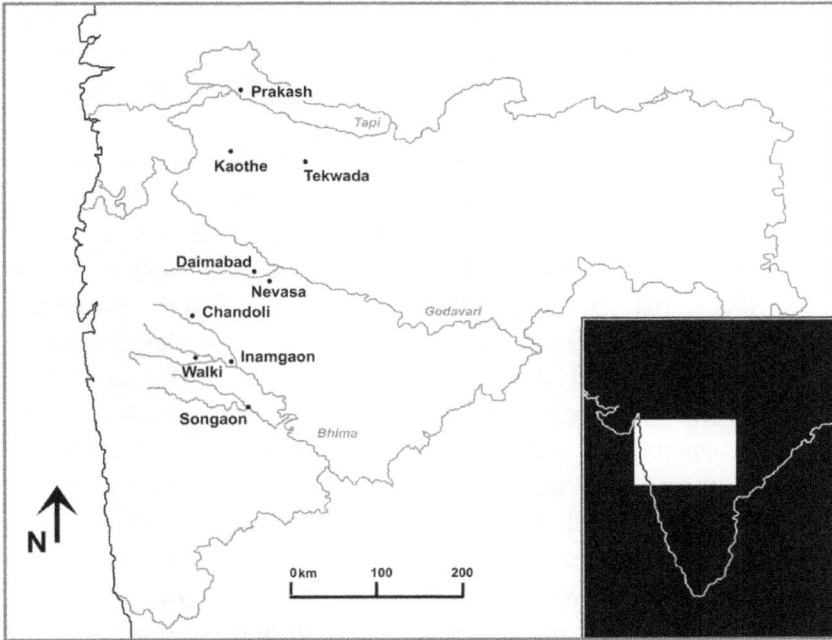

Figure 1.2. Map of Chalcolithic sites in Maharashtra during the second millennium B.C.

2002). The average size of these settlements declined from 13.5 hectares (ha) to 3.6 ha during the post-urban phase.

During the second millennium B.C. agricultural villages also dotted the landscape of Maharashtra (Fig. 1.2). These 'Copper and Stone Age' villages of west-central peninsular India are collectively referred to by the culture history term "Deccan Chalcolithic." Based primarily on ceramic typology, this period is divided into four major phases: the Savalda (2200–1800 B.C.), Late Harappan (1800–1600 B.C.), Malwa (1600–1400 B.C.), and Jorwe (1400–700 B.C.). It was during the Savalda (2000–1800 B.C.) that the first Deccan Chalcolithic settlements appear in northern Maharashtra, in the Tapi and Pravara river valleys. The majority of these villages were small, covering about 1 ha of land, and population size in an average village was approximately 100 to 200 people[1] (Dhavalikar 1988). During this time, Deccan Chalcolithic people built round or rectangular mud and mud-brick houses. They practiced a diverse subsistence economy based on hunting, fishing, foraging, stock raising, and subsistence agriculture

(Dhavalikar 1988; Dhavalikar, Sankalia, and Ansari 1988; Shinde 2002). They had well-built ceramics and copper technology, although the use of copper was sporadic, perhaps due to the scarcity of raw materials in this area.

There were about 50 Deccan Chalcolithic settlements in northern Maharashtra by the beginning of the Late Harappan phase (1800–1600 B.C.). A brick-lined burial chamber, bronze figures, and carved inscriptions at Daimabad resembling Harappan seals are evidence that some of these villages maintained contact with Indus trade networks during this time (Sali 1986). Significant population expansion began during the subsequent Malwa phase (1600–1400 B.C.), and the site of Daimabad soon became a regional center within the Pravara River valley. Archaeologists have speculated that population growth was the impetus for Chalcolithic people to incorporate more agricultural production into their lifestyle during the subsequent Jorwe phase (Dhavalikar 1988).

At present, the origin of Deccan Chalcolithic people and their connection to the Harappans remains unclear. The Indus Origin hypothesis suggests that Indus people resettled in central India after the breakup of the Indus civilization, founding villages in what became known as the Malwa and Ahar cultures. Subsequently, through progressive moves southward roughly every 200 years, they eventually colonized Maharashtra (Shinde 1990). Geography, settlement pattern, radiocarbon dates, and some of the structural features at villages assigned to the Malwa culture provide some support for the hypothesis that the Deccan Chalcolithic region was at least in contact with post-urban Indus people. The Malwa region lies east of the Banas Valley and the Aravalli Hills, and south-west of the Ganges-Jumna doab, in an intermediate zone between northwest India and the peninsula. Furthermore, radiocarbon dates suggest some of the settlements were contemporaneous with the post-urban phase of the Indus Age, from 1865–1365 B.C. (Possehl and Rissman 1992). Chalcolithic villages share features with Indus Age settlements, for example, Navdatoli (Sankalia, Deo and Ansari 1971), Nagda (IAR 1956), and Eran (IAR 1961, 1962, 1963, 1964, 1965, 1988) were located on the tributaries of major rivers, the Narmada and Chambal. Navdatoli had a twin village settlement pattern, a feature common to Indus villages (Sankalia, Deo, and Ansari 1971). Similarly, Nagda was a well-planned, organized settlement constructed of mud bricks, resembling standardized brick sizes found at Indus villages (Banerjee, 1986). Ramparts constructed at Nagda and Eran to contain flooding also resemble architectural features common at Indus villages (Dhavalikar 1997), although this feature could be coincidental.

However, radiocarbon dates at Chalcolithic sites in Madhya Pradesh indicate the situation was more complex than the Indus Origins hypothesis suggests. For example, the village of Kayatha shared similarities in artifact styles with Malwa and Ahar material culture and the early, pre-urban Indus village at Kalibangan (Dhavalikar 1997). However, radiocarbon dates demonstrate Kayatha was founded at the height of the Indus civilization, around 2400 B.C., and was abandoned around 2000 B.C. indicating that Indus influences in central Indian villages are not entirely explained by settlement of post-urban Indus refugees (Dhavalikar 1997).

Similarly, radiocarbon dates from villages ascribed to the Ahar culture of Rajasthan revised ideas about the relationship among Ahar, Indus, and Deccan Chalcolithic villages. The Ahar culture of Rajasthan includes more than 100 sites spread over 32,000 km² between 24 and 27 degrees north latitude and 73 to 76 degrees east longitude in the contemporary state of Rajasthan (Hooja 1988), a geographic location between Indus and Deccan Chalcolithic territory. The village of Balathal (24°43' N, 73°59' E) in Rajasthan demonstrates affinities with Indus traditions in the layout of the settlement, construction methods used to build a large stone enclosure at the village, and in ceramic styles (Misra 1997). This village also possesses some defining characteristics of Deccan Chalcolithic culture, such as some similarities in burial traditions; wheel-thrown, decorated ceramics; specialized blade/flake industry; subsistence based on farming, stock raising, and limited hunting; and rectangular houses (Shinde 2000; Robbins et al. 2009). Yet radiocarbon dates from Balathal indicate that sedentary village life in the Copper and Stone Age of Rajasthan had its own indigenous development circa 3400 B.C., contemporaneous with the early Kot Diji phase of the Indus civilization, long before the post-urban phase (Misra 2005). This settlement was occupied until 2000 B.C., after the founding of the first Deccan Chalcolithic villages.

Thus the origin of the Deccan Chalcolithic and the relationships among Indus, Malwa, Kayatha, Ahar, and Deccan Chalcolithic traditions remain uncertain. It seems likely that individual villages in these regions had different relationships with one another and that the relationships were dynamic, changing over the centuries. Many Deccan Chalcolithic traditions may have been a product of indigenous development and culture contact. Material culture of the earliest Savalda villages in the Tapi Valley may represent a tradition with local roots, while the village at Daimabad may have been integrated into an interregional network during the Late Harappan phase. An increase in immigration from Gujarat and central India during that time would explain the more rapid growth rate at that site after 1800 B.C. (Shinde 1984, 1985; Sali 1986; Dhavalikar, Sankalia, and Ansari 1988; Shinde 1989).

What is clear is that additional archaeological research and more radiocarbon dates will be necessary to test hypotheses about the origins of the Deccan Chalcolithic people. To date, fifteen archaeological sites in the Deccan Chalcolithic region have been excavated on a large scale: Prakash, Savalda, Bahurupa, Bahal, Kaothe, and Tekwada in the Tapi basin; Nasik, Jorwe, Daimabad, Nevasa, and Apegaon in the Pravara/Godavari basin; and Chandoli, Walki, Inamgaon, and Songaon in the Bhima/Ghod basin. More than 200 archaeological sites have been recorded in Maharashtra, the majority of which have not been excavated extensively to date. Excavation of these sites will provide valuable insight into the origin and movements of people who lived in the Deccan Chalcolithic region during the first half of the second millennium B.C.

A Climate-Culture Change Model for Understanding the Jorwe Phase of the Deccan Chalcolithic

Research on the Jorwe phase (1400–700 B.C.) of the Deccan Chalcolithic has focused intensely on a totally different set of research questions about climate and culture change, subsistence transition, and their effect on human populations. The Jorwe is divided into two distinct phases named the Early Jorwe (1400–1000 B.C.) and Late Jorwe (1000–700 B.C.). While the Early Jorwe phase is characterized by an expanding population, increases in the number of settlements and their size, and the diversity of site types, after 1000 B.C., most of the region was abandoned. Very few sites persisted into the Late Jorwe phase. Those that were not abandoned saw dramatic shifts in architectural and material culture styles, population density, and subsistence. For decades, archaeologists and bioarchaeologists have sought to understand why.

There are more than 200 archaeological sites in Maharashtra that were occupied during the Early Jorwe phase. These sites range from surface scatters of artifacts and features 0.5 ha in diameter to deep deposits that have created large mounds on the landscape, 30 ha in area. The majority of settlements occupied during the Early Jorwe phase were agricultural hamlets 1–3 ha in area with population sizes between 200 and 600 people (Dhavalikar, 1988). The smaller sites (< 1 ha) include artifact scatters that represent seasonal campsites or locations where occasional pastoral or manufacturing activities occurred. Some of these smaller sites appear to have been farmsteads inhabited only during planting or harvesting periods. It has been suggested that others may have been satellites of larger settled communities, places that were marginalized by social, political, and/or economic processes (Shinde 2002).

The diversity represented by the archaeological record for the Early Jorwe phase of the Deccan Chalcolithic is overshadowed by insights gained from excavations at larger settlements, which are usually considered regional centers (Dhavalikar 1988, 1997; Shinde 2000). Each river valley appears to have had one such center—Prakash in the Tapi basin, Inamgaon in the Bhima basin, and Daimabad in the Godavari basin. Inamgaon and Daimabad are the best understood sites for this time period. From a small settlement founded during the Malwa phase, Inamgaon increased to a size of 5 ha during the Early Jorwe. The population size was estimated to be 1,000 during this time. This site is not only large and well preserved but also the subject of an extensive excavation program through Deccan College Post-Graduate Research Institute, and much has been written about the remains from this site. In addition, the site at Inamgaon is one of only a few sites occupied during the succeeding Late Jorwe phase, and thus it is important for characterizing that transition.

Daimabad grew from 5 ha in the Malwa to 30 ha during the Early Jorwe phase, reaching a population size of approximately 6,000. Both villages, at Inamgaon and Daimabad, were associated with small satellite settlements. Walki, a small campsite near Inamgaon, may have been used as an outpost during the growing season (Dhavalikar, 1994a). Nevasa, a small village across the river from Daimabad, was a small farming village populated by 500 to 1,000 people. The nature of the relationship between Nevasa and Daimabad is unclear, and further research would be required to address that issue.

Settlements occupied during the Early Jorwe phase generally relied heavily on cultivation of drought-resistant barley, wheat, sorghum, and millet. The inhabitants of larger villages, such as Inamgaon and Daimabad, practiced double cropping and may have had incipient irrigation facilities, providing a means for them to grow rice, gram, pulses, beans, and peas

(Kajale 1988; Fuller 2006). The economic system was multi-faceted, featuring agricultural activities supplemented by stock raising cattle, buffalo, sheep, and goats (Thomas 1988; Pawankar 1996); gathering of wild plants, such as Indian jujube (Kajale 1988); deer, antelope, and pig hunting in nearby forested tracts; fishing; and harvesting of freshwater mollusks and other small food items from the local area (Thomas 1988; Pawankar and Thomas 1997). Smaller villages specialized in foraging, fishing, or stock raising throughout the Deccan Chalcolithic and toward the end of the Early Jorwe phase, they increasingly adopted farming at even the smallest, peripheral settlements (Shinde 2002).

This adaptive diversity during the Early Jorwe phase led to relative prosperity and regional stability for a thousand years, with a growing population and a lifestyle well adapted to life in this semi-arid region of west-central India. However, toward the end of the Early Jorwe phase, many Deccan Chalcolithic settlements were abandoned, including the regional center at Daimabad. Inamgaon is the best understood of only a handful of villages that persisted into the Late Jorwe phase (1000–700 B.C.).

Archaeologists have characterized this phase at Inamgaon as a shift from sedentary agricultural life to semi-nomadic pastoralism based on several lines of evidence (Dhavalikar and Possehl 1974; Dhavalikar 1977, 1984; Dhavalikar 1988; Dhavalikar, Sankalia, and Ansari 1988; Dhavalikar 1989, 1994b). Analysis of floral remains (charcoal and charred seeds) shows there was a decline in agricultural production during the Late Jorwe phase (Kajale 1988) and fewer cereals were consumed (Gogte and Kshirsagar 1988). Zooarchaeology, or analysis of faunal remains present in food refuse, demonstrated a shift in emphasis from cattle keeping during the Early Jorwe phase to sheep and goat herding during the Late Jorwe phase (Thomas 1988; Pawankar 1997; Pawankar and Thomas 1997). There was also a shift in house construction from rectangular houses built during the Early Jorwe phase to round houses built during the Late Jorwe phase (Dhavalikar 1988). In Indian ethnoarchaeology, round houses are associated with poverty and increased likelihood of mobility (Allchin 1994).

The archaeological evidence from the Late Jorwe phase has usually been interpreted using a climate-culture change model that proposes an increase in aridity after 1000 B.C. led to the decline of agricultural settlements. The model was developed primarily from paleoclimate research based on pollen samples collected from lake cores in central Rajasthan (Singh 1971; Singh, Joshi, and Singh 1972; Singh et al. 1974) and studies of fluvial dynamics in west-central India (Rajaguru and Kale 1985; Kale and Rajaguru 1987, 1988; Kale 1999; Kale 2002). The data from Rajasthan suggested oscillations between humid and dry phases occurred throughout the Holocene. From this data, a sequence for climate changes over the past 10,000 years was developed (Singh 1971; Singh et al. 1974) and widely applied to the archaeological record across South Asia. The Rajasthan sequences suggested a period of wetter climate from 3000–1700 B.C., more arid conditions from 1700–1500 B.C., a relatively wet phase from 1500–1000 B.C., then arid conditions returning around 1000 B.C. and continuing until 500 B.C.

This paleoclimate reconstruction supported ecological hypotheses for the end of the urban phase of the Indus Age (Wheeler 1959; Fairservis 1979; Allchin and Allchin 1982; Raikes and Dales 1986) because the mature phase of the Indus Age appeared to coincide with a humid phase while the post-urban decentralization roughly coincided with the onset of an arid phase in the Singh reconstruction. This climate reconstruction has been very influential in South Asian archaeology in general, and over the years it served as the foundation for a climate-culture change paradigm that was widely applied to interpret sites across South Asia, including explanations for the abandonment of sites toward the end of the Deccan Chalcolithic period. The following quotes provide a summation of how climate change, inferred from Singh's data, was used to interpret the Late Jorwe phase. In "*The First Farmers of the Deccan*," Dhavalikar (1988) stated,

The aridity which sets in towards the close of the second millennium B.C. seems to have been very severe and this probably was the cause of the downfall of Chalcolithic cultures. Everywhere in Central and Western India, the entire Chalcolithic activity comes as it were to a grinding halt. By the close of the second millennium B.C., the settlements were deserted by the early farming communities to be occupied again only after some five centuries in the beginning of the 6th–5th cent B.C. . . . It would therefore be clear that the principal cause of the downfall of the early farming societies in Central and Western India was the climatic change which ushered in increasing aridity that continued for about five centuries from circa 1000–500 B.C. (33–34)

In 1994, Dhavalikar again summarized the transition:

> It appears that there was a drastic change in climate at the end of the second millennium (Kale and Rajaguru 1987). Successive droughts caused economic decline, as a result of which the Chalcolithic habitations all over central India and the Deccan were deserted by early farming communities. But in the Bhima valley, the farming people survived somehow. They were reduced to abject poverty, which is reflected in their small round huts, coarse pottery, and dwindling agriculture. This culture, labeled Late Jorwe phase, can be dated from c. 1000 to 700 B.C. In time the conditions worsened still and the people gradually resorted to semi-nomadic, sheep and goat pastoralism (Dhavalikar, 1989). They finally deserted the settlement by 700 B.C. when southern megalithic people began their incursions into the northern Deccan with their superior iron weaponry and fast-moving horses. (33)

The climate-culture change model predicted dire consequences for human populations and health with the failure of agriculture at Deccan Chalcolithic sites. This perspective was challenged by bioarchaeologists who generally view agricultural subsistence in a less than positive light. Bioarchaeologists have examined human skeletal material from the Deccan Chalcolithic to address two inter-related research questions: Is there evidence for subsistence transition? What effect did the transition have on human health and stress levels during the Late Jorwe phase at Inamgaon?

A Subsistence Transition Model for Understanding Health in the Early and Late Jorwe Phase

Agriculture was once believed to represent the most significant achievement of human societies prior to the industrial revolution. Scholarly and popular perception long reasoned that agriculture was responsible for increased capacity to produce food, the development of food storage led to improved nutritional status, longevity, improvements in health in general, and increased leisure time, and that these provided a means for ensuing cultural florescence, and even served as the foundation for hierarchical civilizations. Human remains from archaeological sites serve as an important source of archaeological evidence to address questions about human-environment interactions, subsistence practices, and their consequences for human populations.

Ecological environments, cultural circumstances, lifestyles, and behaviors shape human biology and leave their mark on the human skeleton. Bioarchaeology is the study of human biology and culture through the analysis of skeletal populations. A substantial amount of bioarchaeological research has focused on subsistence transition—the effects of agriculture and other dietary changes on human populations (Cohen 1977; Lukacs 1983; Cohen 1984, 1989; Lukacs 1992; Tayles 1992; Lukacs and Pal 1993; Larsen 1995; Tayles 1996; Nelson, Lukacs, and Yule 1999; Lambert 2000; Tayles, Domett, and Nelsen 2000; Pietrusewsky and Douglas 2001; Steckel et al. 2002; Pietrusewsky and Tsang 2003; Hutchinson 2004; Lukacs and Pal 2004; Oxenham, Thuy, and Cuong 2005; Oxenham, Nguyen, and Nguyen 2006; Cohen and Crane-Kramer 2007). With some exceptions for rice agriculturalists (Pietrusewsky and Douglas 2001; Oxenham, Thuy, and Cuong 2005; Oxenham and Tayles 2006; Domett and Tayles 2007), anthropological observations of human societies around the globe suggest systematic reductions in quality of life accompanied agricultural subsistence transitions in many areas of the world, including reduced skeletal and dental health.

Bioarchaeology has demonstrated a predictable impact of agricultural subsistence on pathological profiles for human populations—reduction in the velocity of childhood growth, reduced adult stature, increased prevalence of micro-nutrient deficiencies and protein-energy malnutrition, markers of growth disruption in the bones and teeth, increased prevalence of some infectious diseases, higher frequencies of some non-specific indicators of infection, and declining dental health related to increased frequency of caries (cavities), abscess, and ante-mortem tooth loss.

Bioarchaeologists used subsistence transition theory to make predictions about the effect of culture change on human health during the Early to Late Jorwe transition in India (Lukacs 1997; Lukacs and Walimbe 1998; Lukacs and Walimbe 2000; Lukacs, Nelson, and Walimbe 2001; Lukacs and Walimbe 2005b, 2007a). Lukacs and Walimbe proposed a model that reduction in settlement density and a shift away from agricultural production during the Late Jorwe phase would lead to an *improvement* in health status. Like Dhavalikar's climate-culture change model, the subsistence transition model proposed by Lukacs and Walimbe suggests that culture changes during the Late Jorwe phase were initiated by an increase in aridity. However, the model differs dramatically in the specific predictions for demography and health. According to Dhavalikar's model, the Late Jorwe phase should demonstrate poorer health circumstances than the Early Jorwe phase due to the collapse of the subsistence system initiated by the climate changes. This model predicts increases in infant morbidity and mortality rates during the Late Jorwe phase compared to the Early Jorwe phase. Lukacs and Walimbe's model on the other hand suggests the skeletal population from the Late Jorwe phase at Inamgaon should appear healthier, with lower prevalence of growth disruption and pathological conditions than the Early Jorwe skeletal population due to the reduced reliance on agricultural food items.

Skeletal materials from the Early and Late Jorwe phase have been examined to test hypotheses from the subsistence transition model, and the results were somewhat contradictory. Skeletal remains representing the Early Jorwe phase were recovered from Nevasa (N = 74) and Daimabad (N = 35). Inamgaon yielded skeletal material from both the Early and the Late Jorwe phase (N = 205). The skeletal collections from the Deccan Chalcolithic sites are notable in that there are very few adults present and 90 percent of the individuals present died when they were less than five years of age. This age structure has advantages and disadvantages. The advantages are that it is easier to estimate age at death for young individuals and age can be based on dental development and eruption schedules, which are fairly predictable and have relatively narrow margins for error. Another advantage is that it is possible to examine growth disruption in the teeth and bones, which are growing and maturing early in infancy and childhood. Disruptions to homeostasis, or stressors, are recorded with higher fidelity in a child than in an adult, who has accumulated a lifetime of stress, insults to growth, and periods of recovery.

There are also disadvantages to working with skeletal populations with few adults present. In a sample of children we cannot see the results of these growth trajectories in adulthood. If a sample is skewed toward younger age categories, there is no method to estimate the mean adult stature or the prevalence of pathological conditions in the subset of the population that survived childhood. In short, there are challenges, limitations, and opportunities specific to the Deccan Chalcolithic skeletal populations because of their age structure and this has shaped the analysis of demography and health for these samples.

Skeletal pathological conditions are not common in either the Early or the Late Jorwe series. Periostosis, or evidence for systemic infection, is the most common skeletal pathology in the collection from Inamgaon, but its prevalence is not significantly different between the two phases, occurring in low frequency for both the Early (2/51, 3.9%) and the Late Jorwe phases (6/109, 5.5%) (Lukacs et al. 1986). Two individuals in the Inamgaon series had spina bifida occulta, a congenital defect related to folic acid deficiency during early stages of pregnancy, characterized by an opening at the base of the spinal column. There was one case of trigonocephaly, an inherited developmental disorder in which the two halves of the frontal bone fuse prematurely causing a narrow forehead. There was a reported case of myostitis ossificans circumstcripta, a rare inherited disorder in bone formation now commonly known as fibrodysplasia ossificans progressive. Finally, there was a low prevalence of greenstick fractures in infants less than three years of age. These are fractures where limb bones bend but do not completely break, common in young infants and children with pliable, incompletely mineralized bones. Thus few individuals expressed any sign of skeletal pathology and there were negligible differences in these health indicators through here.

Limb bone lengths for subadults from the Deccan Chalcolithic sites were compared with a modern sample of living children in Maharashtra (Walimbe and Gambhir 1997). Based on their own growth standards, 75 percent of Chalcolithic sub-adults examined (n = 192) had long bone (i.e., limb-bone) lengths less than two standard deviations below the average for age in their contemporary peers (Table 1.1). Individuals in the second year of life (around 18 months old) were significantly more likely to demonstrate growth suppression, which was attributed to weaning stress. This study did not attempt to test for significant differences among the Early and Late Jorwe samples but it provides useful information about diet, nutritional sufficiency, and subadult growth in contemporary Maharashtran villages. The authors also provide ethnographic information about breastfeeding, supplementation timing and foods, information about maternal nutrition, depletion, and inter-birth interval in these communities, which confirms the relationship between socio-sanitation conditions, weaning stress, and kwashiorkor for rural communities in India.

In general, populations that rely on agricultural subsistence have a relatively high frequency of dental caries (commonly known as cavities). This generalization does not always hold true; for example, caries rates

Table 1.1. Results of prior research on growth profiles at Inamgaon (% individuals).

Nutritional grade	Malwa	Early Jorwe Inamgaon	Late Jorwe Inamgaon
Normal		22.22	23.08
I	28.57	16.67	38.46
II	50.00	61.11	34.62
III	14.29		
IV	7.14		3.85
Sample size	14	18	26

Source: Data from Walimbe and Gambhir (1997:118).

are generally higher in populations that rely heavily on maize or cereal crops like barley as opposed to populations relying primarily on rice agriculture (Oxenham, Nguyen, and Nguyen 2006). Populations with higher rates of fertility also tend to have higher frequencies of dental caries due to synergistic relationships among fecundity, oral biology, eating habits, food choices, and dental health (Lukacs 2008). South Asian skeletal populations from north and west India have repeatedly demonstrated a relationship between increased agricultural production and increased caries rates among individuals and among teeth affected (Lukacs and Walimbe 2007a). Inamgaon too demonstrates a statistically significant reduction over time in the proportion of postcanine teeth affected by caries. The Late Jorwe subadult sample had lower caries rates than the Early Jorwe subadult sample. This result supports the hypothesis that the Late Jorwe population was relying less heavily on agriculture and cariogenic foods compared to Early Jorwe population.

Tooth size reduction is also related to dietary processing and can change in response to subsistence transition (Hinton, Smith, and Smith 1980; Kieser 1990; Smith and Horowitz 2007). Odontometry (measures of tooth size) was used to examine the effects of reduced reliance on agriculture during the Late Jorwe phase at Inamgaon, but tooth sizes and crown biting surface areas demonstrated no significant differences in the deciduous dentitions from the Early and Late Jorwe skeletal samples at Inamgaon (Lukacs and Walimbe 2005a; Lukacs and Walimbe 2007a).

Although tooth breadth, length, and areas did not differ significantly in the permanent dentitions, there was one significant difference between the Early and Late Jorwe samples in summed tooth crown area (Lukacs and Walimbe, 2007a). This measure is the sum of areas across the entire dentition and it reflects the amount of occlusal, or chewing, surface available. It is expected that occlusal surface area will be reduced in a population that relies on agricultural subsistence although the precise mechanism of dental reduction is uncertain. The summed crown area for the complete permanent dentitions was greater during the Late Jorwe phase. Permanent teeth from the Early Jorwe phase at Inamgaon had a summed tooth crown area of 1,192 mm², whereas the same measure in the Late Jorwe sample was 1,236 mm². This difference of 44 mm² was considered statistically significant and was used to infer that tooth size increased significantly between the Early and Late Jorwe phases at Inamgaon, perhaps due to an influence from subsistence practices and diet. This result supports the hypothesis that there was a subsistence transition and the Late Jorwe population relied less heavily on cereal foods and agricultural production.

Enamel defects (linear enamel hypoplasia, or LEH; interproximal contact hypoplasia, or IPCH; and localized hypoplasia of the primary canine, or LHPC) represent periods of time when dental development is halted or slowed due to elevated stress levels, or disruptions to homeostasis (Goodman, Armelagos and Rose 1980). These dental stress markers have been studied extensively in available dentitions from Inamgaon, Nevasa, and Daimabad. LEH is an enamel defect that occurs in permanent teeth with very low frequency for hunting and foraging populations but with high frequency for agricultural populations worldwide (Cohen 1984; Lambert 2000; Steckel and Rose 2002). The subsistence transition model predicts that the Early Jorwe skeletal population from Inamgaon will have a higher frequency of LEH because reliance on agriculture was greater, whereas the Late Jorwe skeletal population from Inamgaon will have a relatively low frequency of LEH due to increased dietary diversity and improvements in sanitation accompanying reduced population size. In a comparison of permanent dentitions from the Early and Late Jorwe phases, there was no significant difference in the proportion of individuals affected by LEH. However, there were two significant differences in the number of stress episodes (the average number of defects per tooth). During the Late Jorwe phase, individuals who were affected by LEH demonstrated fewer defects on average in the lower central incisor. However, the upper canine tooth had a higher average number of defects per tooth during the Late Jorwe phase at Inamgaon. Unfortunately, the maxillary canine and the mandibular central incisor develop on different schedules and have demonstrated different levels of sensitivity to disruptions in enamel formations and thus these results do not point to a clear difference in stress levels during the Early and Late Jorwe phases at Inamgaon.

Enamel defects can also be examined in the deciduous dentition (baby or milk teeth). IPCH is a defect that occurs on the interproximal surfaces of the anterior primary teeth. The etiology of this defect is unclear, but Skinner (1996) suggested its cause was related to contact between neighboring teeth during development and Lukacs (1999) found support for this hypothesis in the form of matching lesions on adjacent teeth. A review of the clinical and anthropological literature suggested that these defects are not uncommon (Lukacs 1999). They are strongly associated with low and very low birth weight, their presence is associated with increased susceptibility to dental caries, and the defects are often found in association with other enamel defect types (particularly LHPC), although this relationship was not statistically significant. At Inamgaon, IPCH declines in frequency during the Late Jorwe phase compared to the Early Jorwe phase, but the difference is not statistically significant (Lukacs 1999).

There was a significant difference in the rate of LHPC through time at Inamgaon. LHPC is another plane form defect that occurs on the primary canine and is associated with low socio-economic status and poor nutrition (Skinner and Hung 1989). The frequency of LHPC was higher during the Early Jorwe phase at Inamgaon (47.4%) and lower during the Late Jorwe phase (35.7%) (Lukacs and Walimbe 1998; Lukacs and Walimbe 2000). The frequency of LHPC in Early Jorwe Nevasa (36.4%) and Daimabad (33.3%) resembled the Late Jorwe phase at Inamgaon. Thus the atypical figure is that of the Early Jorwe at Inamgaon (Lukacs, Nelson, and Walimbe 2001). Although these defects have a complex etiology (Lukacs 2009), the elevated frequency of LHPC at Early Jorwe Inamgaon was statistically significant and these results were interpreted as providing support for the subsistence transition model by demonstrating a reduction in developmental stress levels during infancy for the Late Jorwe (Lukacs and Walimbe 2007a).

The subsistence transition model has two parts to it, and an examination of the studies that were conducted on the human remains from Inamgaon reveals two separate but related research questions and conclusions. The first question concerns whether there is evidence for subsistence transition from heavy reliance on cereal foods and agricultural production during the Early Jorwe phase at Inamgaon to reduced reliance on cereals in the diet during the Late Jorwe phase. This half of the model has been confirmed using dental anthropology. Previous studies have demonstrated that the Late Jorwe sample had lower caries rates and demonstrates a significant level of tooth size reduction when compared to the Early Jorwe sample from Inamgaon. This fits with expectations for populations that are not relying heavily on cereals, and these results support the hypothesis that Late Jorwe people had greater dietary breadth.

The other part of the subsistence transition model concerns the *effects* of relying on agriculture during the Early Jorwe phase and reduced agricultural production during the Late Jorwe phase. Archaeologists predicted a decline in health status would accompany the subsistence transition, as it was initiated by a collapse of part of the subsistence system. However, because agriculture is generally associated with negative impacts on human health and elevated biocultural stress levels, comparisons of osseous pathology profiles and dental enamel hypoplasias were conducted to examine the hypothesis that the Early Jorwe was more stressed due to their reliance on cereal agriculture and their sedentary village lifestyle. It was hypothesized that Late Jorwe people were healthier or had a lower frequency of stress markers because they relied less on agriculture. The results of these studies have been much less conclusive. Osseous pathological profiles and dental defects (LEH and IPCH) demonstrate no conclusive trend in regard to stress levels. These results suggest that the Early and the Late Jorwe phases had broadly similar stress levels. These markers do not support the hypotheses developed from the subsistence transition model. On the other hand, examination of LHPC rates appears to confirm hypotheses developed from the subsistence transition model. Rates of LHPC were high in the Early Jorwe sample at Inamgaon compared to the lower rates found at Late Jorwe Inamgaon, Nevasa, and Daimabad.

Unfortunately, none of these studies have been unable to account for the "osteological paradox." The frequency of biocultural stress markers in a skeletal population can be affected by its age structure and its demographic profile, fertility, and infant mortality rates (Wood et al. 1992; Katzenberg and Saunders 2000; Wright and Yoder 2003; Lukacs 2008). By definition, skeletal populations are comprised of individuals who died and so they are not a perfect representation of what the living population looked like. Low frequency of developmental stress markers can occur in a subadult skeletal population because of low stress levels or it can occur due to high infant mortality rates—because infants died before they could express the effects of chronic conditions or growth disruption in the skeleton and teeth. Similarly, high frequencies of stress markers in a skeletal population can result from elevated stress levels or from a larger number of "healthy survivors." Paradoxically, an apparently "healthy" skeletal population may actually represent high levels of stress in a population and vice versa. On the other hand, it is possible to compare relative stress levels among skeletal samples using converging lines of evidence from demography, skeletal and dental morbidity, and acute and chronic stress markers (Goodman and Martin 2002; Cohen and Crane-Kramer 2007). For this reason, bioarchaeologists must evaluate pathological profiles within the context of information about demography.

Until we have a model for fertility and infant mortality rates at Deccan Chalcolithic villages, it is not possible to interpret biocultural stress markers with much certainty. Previous work had begun an examination of demography in the Deccan Chalcolithic (Lukacs 1980; Lukacs and Badam 1981; Lukacs and Walimbe 1984); however, this work suffered from two serious but common methodological issues (see Sattenspiel and Harpending 1983; Jackes 1992; Meindl and Russell 1998; Paine and Harpending 1998; Wright and Yoder 2003). Traditional mortality-centered approaches to demography rely on an assumption of nonstationarity, but we know that Jorwe populations were not stationary—there was population growth during the Early Jorwe phase and a decline in population size during the Late Jorwe phase. We also know that fertility has a large impact on demographic structure, particularly when a population is growing or declining in size. Unfortunately, it was not possible to apply a fertility centered approach to demography at Inamgaon, Nevasa, and Daimabad because of the age structure of the assemblages. Fertility centered methods for paleodemography traditionally focus on proportions of older children and adults (Bocquet-Appel and Masset 1982; Buikstra, Konigsberg, and Bullington 1986; McCaa 1998, 2002).

A Biodemographic Model for Climate and Culture Change During the Deccan Chalcolithic

In this book, I will provide evidence from recent paleoclimate studies and examine the hypothesis that there was climate change toward the end of the Early Jorwe phase of the Deccan Chalcolithic. It is necessary to synthesize recent developments in paleoclimate research as the majority of studies have relied on the reconstructions developed by Singh and colleagues during the 1970's. In chapter 2, I will demonstrate there is no conclusive evidence for dramatic climate changes at the end of the second millennium B.C. in India. Recent paleoclimate evidence demonstrates a semi-arid climate was already established in west-central India by 2000 B.C., at the beginning of the Deccan Chalcolithic. Many of the contemporary reconstructions suggest that aridity was increasing throughout the second millennium, but it actually peaked prior to the Early Jorwe phase, not at its end.

This book will also characterize the archaeological record of the Deccan Chalcolithic at Inamgaon, Nevasa, and Daimabad. In chapter 3, I will present archaeological evidence on the biocultural adaptations of Early Jorwe people. Their strategies for dealing with the semi-arid climate of peninsular India included a diverse economy and heavy reliance on drought-resistant crops, such as barley. The Early Jorwe phase was characterized by relatively rapid expansion of population sizes and increased diversity in the types of communities in the Deccan region. During the Late Jorwe phase, the archaeological record at Inamgaon demonstrates significant changes to the subsistence activities at Inamgaon. The people moved away from the focus on barley production. The archaeological evidence suggests that there was a shift in species preferences throughout all categories of subsistence activity and that the total amount of food production declined significantly. Saline-tolerant species became preferred for agriculture. Freshwater shellfish replaced other protein sources gathered during the Late Jorwe phase. Foods that are commonly eaten during periods of starvation in India today began to be utilized with increasing frequency during the Late Jorwe phase. Archaeologists have also demonstrated declines in socio-sanitation conditions due to partial abandonment of the settlement and accumulating refuse, including burials under the abandoned house floors. Thus the archaeological record clearly indicates a decline in socio-sanitation conditions and local environmental degredation accompanied the abandonment of many villages. Although the occupation at Inamgaon persisted, there were significant changes in the culture, subsistence, and quality of life during the Late Jorwe phase.

Based on the evidence summarized in chapters 2 and 3, I will evaluate an alternative biodemographic model for culture change during the Deccan Chalcolithic. I suggest Early Jorwe populations were highly successful in the sense that they were enjoying rapid population growth and huge increases in settlement density. However, over the course of 400 years of settled village life, growing populations became increasingly dependent upon drought-resistant barley agriculture to supplement hunting and foraging activities. The large population sizes and increasingly unsustainable agricultural practices eventually resulted in local resource depletion, local ecological degradation, soil salinization, and socio-sanitation issues in Deccan Chalcolithic villages. Large population centers like Daimabad did not adjust their behavior in time, and a lack of flexibility may have contributed to their collapse. Small satellite settlements like Nevasa soon followed. Despite poverty, food shortages, poor sanitation, and declining health status, mid-sized settlements like Inamgaon retained enough flexibility to persist for another 300 years. The Late Jorwe population at Inamgaon responded to local environmental degradation by relying on a new mix of saline-tolerant crops, herding, increased foraging, and hunting activity. However, the Late Jorwe population was stressed, and eventually the small, stressed population also collapsed.

This model generates specific expectations for demography and human health profiles. Biodemography is an integrated approach to bioarchaeology that recognizes a link between population growth (settlement density, growth rates, fertility, and mortality) and growth of the body (height, body mass, and limb bone lengths). My overriding concern is to characterize the transition at Inamgaon within the larger context of variation at Early Jorwe sites as much as possible. To accomplish that goal, I will construct demographic profiles and compare estimates of population dynamics for Nevasa, Daimabad, and the Early and Late Jorwe at Inamgaon. My first hypothesis is that demographic profiles for Early Jorwe settlements will demonstrate heterogeneity, with differences in fertility and infant mortality corresponding to differences in settlement size, density, and growth rate. To evaluate this hypothesis, I will examine the demographic profiles from Early Jorwe sites along an urban-rural continuum: the large growing settlement at Daimabad, the moderately sized and slower growing village at Inamgaon, and the smaller stable settlement at Nevasa. The biodemography model predicts that during the Early Jorwe phase, the large growing settlement at Daimabad, a regional center, will demonstrate demographic dynamics significantly different from other, smaller Early Jorwe settlements (Nevasa and Inamgaon). The variation present in the Early Jorwe phase profiles will then serve as a baseline with which to compare the Late Jorwe phase at Inamgaon and to characterize the nature of the Early to

Late Jorwe transition to understand why so many settlements were eventually abandoned.

My second hypothesis is that the demographic profile for the Late Jorwe phase at Inamgaon will demonstrate significant differences from the Early Jorwe phase at Inamgaon. Different models of culture change at Inamgaon lead to different predictions for the direction of those differences between the demographic profiles. Dhavalikar's climate-culture change model has two parts: climate change and culture change. The climate change portion of his model will be addressed by paleoclimate evidence; demography can address the nature of the culture changes. Dhavalikar suggested that Late Jorwe people lived in impoverished and "dire" circumstances after the collapse of the subsistence base (Dhavalikar 1988). Alternatively, Lukacs and Walimbe's model suggested there was an improvement in circumstances when agriculture was abandoned during the Late Jorwe phase due to increased dietary diversity, less reliance on cereals, and reductions in population size and density. The demographic situation is described in chapter 4. Fertility and infant mortality rates were estimated for Deccan Chalcolithic populations based on a new technique for fertility-centered demography in subadult samples (Robbins, in press). These new estimates shed light on whether the decline of Chalcolithic settlements occurred in a high or low pressure demographic transition.

The prevalence of biocultural stress markers should be interpreted based on differences in demography, and I will integrate this information into my interpretation of childhood growth and development in the Jorwe phase. I analyze growth at Inamgaon using long bone lengths, stature, body mass, and body mass index, or BMI (body mass scaled to estimates for stature squared). This approach is in part derived from a new technique for estimating body mass in subadult skeletons using compact bone geometry (Robbins, Sciulli, and Blatt 2010), which is explained in chapter 5. Results of the analysis are provided in chapter 6. I expect rates of emaciation and skeletal growth suppression will be significantly different over time at Inamgaon based on the significant cultural and subsistence changes that occurred. The biodemography model suggests that the growth of individuals and the growth of populations are intertwined. Thus changes in growth profiles are influenced by changes in demographic dynamics. I predict that mortality rates will change through time at Inamgaon and rates of growth disruption in the skeleton will also differ significantly between the Early and the Late Jorwe at Inamgaon. If

Late Jorwe Inamgaon is characterized by a high-pressure demographic situation (relatively high fertility and low life expectancy at birth) and this skeletal population demonstrates increased rates of emaciation, this result would indicate support for the dire consequences suggested by Dhavalikar's climate culture change model. If declines in settlement density during the Late Jorwe are accompanied by a relatively low -pressure demographic situation (lower fertility rates and higher life expectancy at birth) and this skeletal population demonstrates decreased rates of emaciation, this result would indicate support for the subsistence transition model—the connection between reduced reliance on agriculture and improvements in health status. If the Late Jorwe is characterized by relatively low life expectancy but the skeletal population also demonstrates low rates of emaciation, this result could indicate that the osteological paradox has affected the pathological profile (see chapter 6 for an explanation).

Broader Issues

The effects of maize agriculture on human populations in the New World are relatively well documented. However, subsistence transition in East Asia is not as well documented and research is increasingly focused on this area (Pietrusewsky 1974; Lukacs 1992; Lukacs and Minderman 1992; Lukacs and Pal 1992; Tayles 1992; Lukacs and Pal 1993; Tayles 1996; Pietrusewsky, Douglas, and Ikehara-Quebral 1997; Lukacs and Walimbe 2000; Tayles, Domett, and Nelsen 2000; Pietrusewsky and Douglas 2001; Lukacs 2002; Pietrusewsky and Tsang 2003; Domett and Tayles 2006; Douglas 2006; Oxenham, Nguyen, and Nguyen 2006; Oxenham and Tayles 2006; Bazarsad 2007; Douglas and Pietrusewsky 2007; Krigbaum 2007; Pechenkina, Benfer, and Ma 2007; Halcrow and Tayles 2008; Halcrow, Tayles, and Livingstone 2008). Despite the wealth of archaeological and biological evidence available from South Asia, this region is often overlooked or included as a footnote in paleoanthropological and archaeological research (Morrison 2002; Lukacs and Walimbe 2007b). It has been suggested that this relative neglect is due to difficulties of dating and stratigraphy, preservation issues, or the distinctive nature of the record, which does not fit easily into models constructed for use elsewhere in the world.

However, the archaeological record for South Asia provides an important opportunity to understand the diversity of human strategies for dealing with ecological changes. The Deccan Chalcolithic spans large geographical and temporal spaces. As communities settled throughout central-west India, they adjusted to changes in access to raw materials for tool production, diverse micro-climates, soil conditions, and availability of wild foods. Although there are some common features of these settlements, such as ceramic typology, food species preferences, and burial practices, there is also a wide range of diversity. In fact, adaptive diversity is one of the many behavioral patterns that arose in response to ecological and social change—populations maintained flexibility in their subsistence practices through a low-level commitment to agricultural production and willingness to shift strategies when circumstances demanded. My intention in this book is to reconstruct some of the diverse strategies employed by three of these communities (Inamgaon, Daimabad, and Nevasa) and to document some of the heterogeneity that may have been a crucial strategy for survival in the semi-arid climate of the Early Jorwe phase.

South Asia is often perceived through a lens of climate, geography, and ecological circumstance. Previous bioarchaeological research on the Deccan Chalcolithic has attempted to understand how biocultural stress levels changed in response to changing ecological and subsistence environments. The novelty of my approach is in the synthesis of a large body of recent paleoclimate evidence, application of new techniques for paleodemography for subadult samples, a new approach to examining subadult growth in skeletal populations using body mass scaled to stature, and my interpretation of the profiles using biodemography. My hope is that this project provides new insights into the short and long term strategies used by past populations for dealing with life in the semi-arid climate of India. More than 500,000,000 million people live in villages in semi-arid regions of contemporary India, and they stand to lose the most in the context of overpopulation, environmental degradation, and global climate change. Bioarchaeology can provide a unique perspective on human adaptations, diet, lifestyle, and epidemiology. The questions are pertinent to modern human situations, global health programs, development projects, and climate change initiatives. The answers are found in the outcomes of the past.

Organization

West-central India is a dynamic landscape offering unique challenges and opportunities for biocultural adaptation. Chapter 2 begins with a discussion of the landscape, the present ecological context, and evidence for climate change during the Holocene. Chapter 3 describes archaeological evidence for Chalcolithic subsistence and responses to ecological challenges—shifting species preferences, and shifting effort from farming to stock raising, hunting, fishing, and gathering. In chapter 4, I provide details on age estimation methods and outline a new method of estimating demographic parameters for subadult samples. In this chapter, I also provide a comparative demographic analysis for Daimabad, Nevasa, and Inamgoan. In chapter 5, I provide an analysis of childhood growth profiles—growth in height and body mass. The chapter begins with an analysis of the relationship between bone mass and body mass in contemporary children and then I provide a technique for evaluating body mass for height in subadult skeletons. I interpret the results of a comparison of body mass for height (BMI) estimates for the children from the Early and the Late Jorwe phase at Inamgaon in light of the demographic profiles and paleoclimate data. In chapter 7, I review my conclusions about the effects of climate and culture change on prehistoric populations in India.

2

◇◇◇◇◇◇◇◇◇◇◇◇◇◇◇

The Western Deccan Plateau

Environment and Climate

The winds rage, the lightning shoots through the air, the herbs sprout forth
from the Earth, the heavens overflow, refreshment is borne to all creatures,
when Parjanya blesses the Earth with rain.

Rig Veda, verse 83 (Buhler 1859)

Diverse ecological circumstances, combined with the forces of culture
and history, have led to tremendous diversity in India's human populations throughout prehistory (see overviews in Allchin and Allchin 1982;
Kennedy 2000; Chakrabarti 2006). India is a broad, triangular peninsula
positioned between 8 and 37 degrees north latitude and 68 and 97 degrees
east longitude. Except for some narrow mountain passes to the northeast
and west, India is bounded on all sides by mountains and sea. The na-
tion of India covers a large area of more than three million kilometers,
territory which includes many different environments, each presenting
a unique challenge for human settlement. Altitude ranges from 8,500 m
above sea level in the Himalayas to 2 m below sea level in the estuaries
of Kerala in southern India. Average annual rainfall can also be extreme,
from 120 m in the Khasi Hills of eastern India to 120 mm at the western
margin of the Thar Desert. In the Thar Desert alone, the average annual
temperature ranges from 0° Celsius in the winter to 50° Celsius in summer. In the words of Duncan (1880), "India presents all the physical features of a vast continent, and in wonderful variety, from the swamps and
forests of the Tarai to the parched up desert of the Thar, from the richly
fertile soil of Bengal to the salt wastes of Kach, from the dull level of the
plains to the lovely scenery of the Western Ghats, and from the low-lying
Carnatic (Karnatik) to the snow clad peaks of the Himalaya" (Duncan
1880).

The most notable feature of the peninsular landscape is the Sahyadri range (or Western Ghats), which rises 700–1,400 m above sea level and serves as an escarpment between two ecozones—the wet rain forests along the narrow Konkan coastal plain and the broad, flat semiarid region east of the mountains known as the Deccan Plateau (Ollier and Powar 1985). The Konkan Coast runs along the western edge of the Indian peninsula. It is a thin strip of land less than 130 km wide situated between the Arabian Sea and the wave-cut terraces of the Sahyadri Mountains. Interspersed along the steep cliffs and terraces of the coastline are ecosystems dominated by mangrove trees (*Rhizophoracea* and *Avicennia)*, which thrive in each embayment, lagoon, marsh, swamp, and estuary. These habitats are characterized by variation in salinity and moisture, high temperature, and high annual rainfall.

The Western Ghats traverse six contemporary states, from Gujarat south to Tamil Nadu, and are a biodiversity hotspot, home to more than 2,000 plant species, 84 species of fish, 87 amphibian species, 89 reptiles, 15 species of birds, and 12 mammalian species (Daniels 1997). Two major drainage systems in the western peninsula originate here (Rajaguru and Kale 1985; Kale and Rajaguru 1987, 1988). A coastal system flows west from the Ghats to drain the Konkan region. The eastern watershed rivers originate in the high-rainfall zones on the eastern side of the Ghats and follow a gradual declivity, moving out onto the plateau in parallel channels that form along dips in the Deccan trap or basement rifts. For most of the year, these are low-energy systems with low channel gradient, large catchment areas, and elongated forms. The rivers hit their peak flow near the end of the southwest monsoon in September or November. These river courses occur primarily within a rain shadow and thus they can dry up completely in early summer.

The Deccan Plateau lies east of the Ghats. It is a land of little topographical relief comprised of rolling plains and occasionally interrupted by flat hills and basalt terraces. Beneath the erosional landscape of the plateau is the Deccan trap (named by Sykes 1833; cited in Pascoe 1964), a block of dark green to black vesicular amygdaloidal and compact basalts laid down approximately 65 million years ago (Partridge et al. 1996; Ravizza and Peucker-Ehrenbrink 2003) during massive eruptions that took place over a period of 10,000 years (Courtillot and McClinton 2002). The rock flow lies in a continuous, horizontal layer over most of west-central India, south of the Vindhyas from Mumbai to Nagpur (Rajaguru 1988) and is uninterrupted but for bright veins of chalcedony, a microcrystalline type of quartz that includes onyx, jasper, and agate.

Most of the Deccan region is covered by either dry deciduous forest in the west or semiarid grassland and thorn scrub in the east. The dry deciduous forest of the western half of the plateau is comprised of numerous tree species, including those useful for food, fodder, apiculture, medicine, firewood, carving, and building. These include *Acacia catechu* (mimosa, or *khair*), *Anogeissus latifolia* (axlewood, or *dhawa*), *Boswellia serrata* (Indian frankincense, or *saler*), *Cassia fistula* (Indian laburnum, or *bandarlathi*), *Dalbergia latifolia* (Indian rosewood, or *beete*), *Diospyros melanoxylon* (ebony, or *kendu*), *Hardwickia binata* (*anjan*), *Mangifera indica* (mango, or *aam*), *Phyllanthus emblica* (Indian gooseberry, or *amla*), *Psidium guvava* (guava, or *amrud*), *Santalum album* (sandalwood, or *chandal*), *Shorea robusta* (*sal*), *Stereospermum personatum* (yellow snake tree, or *patar*), *Syzigium jambolanum* (plum, or *jamun*), *Tamerindus indica* (tamarind, or *imli*), and *Terminalia paniculata* (Indian laurel, or *sadar*) (Meher-Homji 1989; Puri, Gupta, and Meher-Homji 1989). Over 700 edible plant species have been documented in these dry forests (Vishnu-Mittre 1981), including 383 fruiting species, 250 species with edible leaves and shoots, 110 species of nuts and kernels, 95 species of edible rhizomes, and 46 species of edible flowers and buds (Mehra and Arora 1985; Mehra 1999).

This area is relatively poor in aquatic resources today, but local Bhil people use cast nets (*sikadi*), gill nets (*tangada*), and dragnets (*bichori*) to exploit several local species of fish, including *Mustus villatus* (*tangra*), *Myatus cavasium* (*shingadi*), *Glossotium giuris* (*khashi*), and *Garra mulya* (*malya*) (Shinde 1984). There are 260 species of birds and 75 species of mammals in this ecozone, including *Elephas maximus* (Asian elephant), *Panthera tigris* (tiger), *Panthera pardus* (leopard), *Felis caracal* (caracal), *Felis lynx* (lynx), *Hyaena hyaena* (hyena), *Sus scrofa* (wild pig), *Cuon alpinus* (wild dog), *Melursus ursinus* (sloth bear), *Antelope cervicapra* (black buck), *Axis axis* (spotted deer, or *chital*), *Tetracerus quadricornis* (four-horned antelope), *Cervus unicolor* (sambar), *Bos gaurus* (gaur), *Hespetia edwards* (mongoose), *Lepus negricola* (hare), and *Ratufa macruora* (grizzled giant squirrel) (Pawankar and Thomas 1997).

The majority of people living in Maharashtra today rely on rain-fed agriculture for subsistence. The success of agricultural activities in the Deccan region is likely a result of the black cotton soil (BCS), a prominent feature of the landscape (Dhavalikar 1984, 1985; Dhavalikar 1988; Raja-guru 1988; Dhavalikar 1989, 1994b, 1997). BCS covers approximately 20 percent of the Indian subcontinent, including large tracts of land in the Deccan Chalcolithic region of western Maharashtra (Dasog 2002). The depth of this erosional soil ranges from 10–30 cm. It has a $CaCO_3$ content of less than 4 percent, and phosphates ranging from 11 to 14.29 mgs (Shinde 1984). It has at least three features ideal for agriculture: the soil is rich in organics (7–9%), is moisture retentive, and requires little plowing. The combination of the soil's characteristics with the hot, dry summer climate and seasonal rainfall results in pedoturbation, whereby the soil forms deep cracks followed by swelling that mixes the surface material and the subsoil (Dasog 2002).

Despite its fertile potential, agricultural productivity in this region today is constrained by erratic rainfall, insufficient monsoon cycles, and persistent drought. Rainfall averages 250–1,150 mm per annum east of the Ghats (Joshi and Kale 1997), falling almost exclusively within the southwest monsoon season. The farmers face a continuous gamble on rainfall in this semiarid region, and yet most farms are small family holdings that are unlikely to have large-scale irrigation facilities. The percentage of agricultural land cultivated under irrigation has increased from 18 percent in 1950 to 43 percent in 2004 (India 2004), but coverage is still uneven in different regions of the country, with irrigation systems available for only 12 percent of cultivated land in Maharashtra (Vidarbha Irrigation 2009).

Thus the monsoon is a powerful force in South Asian agriculture. Farmers rely on accurate predictions of monsoon to time the sewing of crops. The rain's predictability, timing, intensity, quantity, surface evaporation, soil erosion, and flooding ultimately determine food sufficiency. Insufficient rain and variations that occur as the storms move across the subcontinent and in weather patterns from year to year are the primary causes of poor timing in sewing the crops, crop failure, food scarcity, and famine (www.maharashtra.gov.in). The major famines of 1865, 1877, 1889, and 1918, for example, were principally a result of insufficient and ill-timed rainfall (Das 1995).

The relationship between the monsoon and agriculture is complex, and each crop has its own specific requirements. *Pennisetum typhoidea* (millet, or *bajra*) requires good rainfall in June, *Sorghum vulgare* (barley, or *jowar*) requires poor rain in June but good rain in July and August, and *Tritium sativum* (wheat, or *rawa*) depends on rain in late September

or October (Panja 1996). Farmers use mixed cropping and double cropping in the *kharif* (summer) and *rabi* (winter) seasons to cope with the
extremely unpredictable and variable nature of the monsoon and the specific requirements of the seeds. Common crops cultivated in both *kharif*
and *rabi* seasons include drought-resistant *Sorghum vulgare*, *Allium cepa*
(onion, or *piaz*), *Solanum indicus* (eggplant, or *brinjal*), and *Hibiscus es-
culentes* (okra, or *bhendi*). *Tritium sativum* and *Raphinus sativus* (radish,
or *muli*) are grown exclusively in the *rabi* season, while *Pennisetum typhoidea*, *Zea mays* (corn, or *makka*), and many varieties of peas are grown
in the *kharif* season. The paleoclimate records for Maharashtra suggest
uncertainty and fluctuations in the amount and predictability of monsoon
rainfall that has affected farmers for millennia, and the archaeological
record indicates coping strategies have changed significantly during that
time (Sankalia, Deo, and Ansari 1960; Shinde 1984; Sali 1986; Dhavalikar
1988; Kajale 1988; Fuller 2006).

Monsoon and Paleoclimate

Over 75 percent of the rainfall on the Deccan Plateau occurs in the southwest monsoon season during the summer months, when the subcontinent
heats up and this low-pressure system pulls wind and moisture in from
the southwestern Arabian Sea. Monsoon rainfall begins at the southwestern tip of the peninsula in the month of June and slowly moves north,
watering the subcontinent until September. One branch of the southwest
monsoon travels northward from Kerala bringing heavy rainfall that wa-
ters the coastal margin and the Ghats. The mountains have an orographic
effect that limits the amount of rainfall on the plateau. In winter the situation reverses. High pressure over the Tibetan Plateau forces cold, dry air
down the slopes of the Himalayan foothills onto the subcontinent and out
to the Arabian Sea, where the air remains relatively cool from December
until March.

The Asian monsoon system is highly variable, in part due to global
processes. Natural cycles of obliquity and precession in the earth's orbit
alter the position of the earth relative to the sun. These cyclical changes
result in glacial and interglacial periods, which are in turn associated with
changes in seasonality in general and the strength of the monsoon system
in Asia (Clemens, Murray, and Prell 1996). Plate tectonics also played a
role in the initial development of the monsoon system, which formed

Table 2.1. Paleoclimatic evidence for the Late Holocene in South Asia.

Citation	Evidence	Date	Climatic inference
Kale and Rajaguru 1987; Rajaguru 1988; Enzel et al. 1990; Jain and Tandon 2003	Riverbank sequence stratigraphy (Maharashtra, Gujarat, and Rajasthan)	3000–700 B.C.	Arid phase
Bryson 1981; Prasad et al. 1998; Roy et al. 2008; Singh et al. 1972, 1973, 1974, 1990; Sinha et al. 2006	Lake sedimentology, palynology (Rajasthan)	3000 B.C.–C.E.	Arid phase in Thar desert; complete dessication of lakes after 3000 B.C.
Prasad and Enzel 2006	Isotopes from lake cores (Rajasthan)	5000 B.C.	Arid phase commenced
Kumaran et al. 2005	Cores from coastal estuaries (Maharashtra)	2500 B.C.	Arid phase began
Caratini 1991, 1994	Deep sea cores (Arabian Sea)	2000–1000 B.C.	Arid phase
Phadtare 2000	Pollen (Garhwal Himalayan region)	2000–1000 B.C.	Arid phase, minimum rainfall around 1500 B.C.
Von Rad et al. 1999	Varves in Pakistan	3000 B.C.–C.E.	Increasingly arid 3000–1900 B.C., maximum aridity 1900–1500 in South Asia and West Asia. Precipitation minimum at 200 B.C.–A.D. 100
Sharma 2004; Sharma et al. 2004b	Pollen, isotopes (Ganga Plains)	3000–1000 B.C.	Arid phase peaked 1000 B.C.

millions of years ago when the Indian subcontinent began subducting beneath the Eurasian plate (Wright 1993).

Although the Holocene in general is an epoch associated with mild climate when compared to the Pleistocene, small shifts in monsoon rainfall, humidity, temperature, sea-level changes, and local ecology have had a significant impact on human populations in South Asia (Yasuda et al. 2004). Paleoclimate is reconstructed using geological, biological, and glaciological sources of evidence (Bradley 1999) and sometimes through interpretation of historical evidence (Dhavalikar 2004). Climate change on a global scale is recorded in deep sea and ice cores. Regionally specific information for South Asia has been derived from riverbank stratigraphy, lake sedimentology, palynology, oxygen and carbon isotope rations, paleopedology, and archaeology. Different techniques provide climate information on a different scale, each technique having a somewhat different level of precision. For example, lake sediments, palynology, and isotope data provide predictions accurate to a period of years, sometimes down to one human generation; whereas paleosols, eolian soils, riverbank stratigraphy, and floral and faunal abundance provide information to make predictions covering longer spans of time, perhaps down to a period of 100 years. This book will focus on diverse sources of evidence about climate in the Late Holocene—after 2500 B.C. (Table 2.1).

Paleoclimate Predictions from Studies of Riverbank Sequence Stratigraphy

Much of the information about paleoclimate in South Asia has been derived from an examination of rivers—their courses, how they change through time, and the deposits they leave behind (Rajaguru and Kale 1985; Kale and Rajaguru 1987, 1988; Rajaguru 1988; Jain and Tandon 2003). In arid or semiarid regions, small fluctuations in climate and local environmental changes can alter the characteristics of a river's channel, significantly impacting flow and sediment load.

In India, shifts in fluvial dynamics are generally attributed to changes in monsoon rainfall. For the endogenic rivers of Maharashtra, which originate in the Western Ghats, a more humid or wetter phase is expected to coincide with a period of riverbank incision and stratigraphic deposits of gravel or gravel-sand bedload (aggradation). Conversely, aridity is expected to leave evidence of ephemeral sand-bed formation due to slower peak flow and increased silt load, a result of the decrease in vegetation along the banks. Sand-bed formation is inferred from an increased supply of sand-sized sediment in the stratigraphy.

Riverbank sequence stratigraphy has frequently been used to reconstruct climate based on examination of layers of soil along incised riverbanks in Maharashtra. Analysis of sequence stratigraphy along the banks of Maharashtra's river valleys in the Chalcolithic period indicates the semiarid climate in the Deccan region was established before 2200 B.C., prior to the Chalcolithic period. This arid climate phase persisted throughout the second millennium B.C., reaching its peak in the Early Jorwe phase of the Chalcolithic (1500–1000 B.C.). During excavations at Daimabad, shallow lenses of fine silt were recorded in the stratigraphy along the Pravara River, indicating to archaeologists that low-energy floods were relatively common in the Early Jorwe phase (Sali 1986). Occasional lenses of pebble grade aggradation in the sequence were interpreted as evidence for high-intensity deluges that would have affected the villages located along this river, including Daimabad and Nevasa (Kale and Rajaguru 1987). This periodic flooding has been implicated as the rationale for locating Chalcolithic settlements slightly away from the river courses, despite the presence of deep, fertile soil and forest cover associated with the banks (Shinde 2002). Low-energy silt deposition and occasional small pebble aggradations were also described along the banks of the Ghod River during excavations at Inamgaon (Rajaguru 1988). These deposits were interpreted as evidence for increased silt availability in the Early Jorwe phase, which could indicate either a reduction in rainfall, or a reduction in the amount of vegetation along the riverbanks, or both.

Research on ancient soil characteristics and formation processes supports the interpretation that semiarid climate was long established in this region. An analysis of paleosols from Inamgaon, Chandoli, and Songaon demonstrated colluvial deposits of basaltic soil formed in semiarid conditions throughout most of the Holocene (Joshi and Kale 1997). Pedocalcic soils that formed after 1100 B.C. have been interpreted as evidence for some level of increasing aridity (Kale and Rajaguru 1987); however, others have suggested that this lens is indicative of soil degradation due to over-irrigation (Kajale, Badam, and Rajaguru 1976; Kajale 1988).

Studies that use sequence stratigraphy to infer fluvial dynamics or reconstruct climate using paleosols provide a broad indication of climate and climate change (Jain and Tandon 2003), but the limits of this approach for understanding large-scale processes are exemplified when rivers in the Deccan region do not demonstrate a profile identical to rivers elsewhere in India—in the Gangetic basin, for example (Kale 2007). Differences between these two regions are evident and could indicate that the monsoon was not performing evenly across the subcontinent (Staubwasser and Weiss 2006), or differences in the stratigraphic profiles of the two regions may indicate that tangential environmental processes affected sediment availability, supply, and base level (Schumm 1993). Maharashtra's rivers lie in a rain shadow, and small-scale local fluctuations in ecological circumstances at the headwaters or anywhere along the river course can have an effect on fluvial dynamics and sedimentation rates (Kale 2007). The human presence was increasing in western Maharashtra during the second millennium B.C., and thus river sediment load could also have been impacted by deforestation, overgrazing of livestock, denudation, or any human activity that increases the sediment load (Rajaguru 1988). Thus ideally these studies will be supported by other lines of evidence from additional sources.

Paleoclimate Evidence from Lake Cores

Reconstructions of paleoclimate in South Asian prehistory have also been developed using a series of lake cores obtained from the northeastern Thar Desert (Singh, Joshi, and Singh 1972; Singh, Chopra, and Singh 1973; Singh et al. 1974; Bryson and Swain 1981; Swain, Kutzbach, and Hastenrath 1983; Wasson, Smith, and Agarwal 1984; Enzel et al. 1990; Singh, Wasson, and Agrawal 1990; Roberts and Wright 1993; Prasad and Enzel 2006; Sinha et al. 2006; Roy et al. 2008). Several techniques have been applied to the cores, including examination of sedimentary facies, carbon and oxygen isotopes, mineralogy, and palynology. Sedimentological studies indicate a Late Pleistocene arid phase during the Last Glacial Maximum.[1] Wetter conditions commenced after that time, and many of the playas of Rajasthan formed toward the beginning of the Holocene. Arid phases punctuate the record for the past 10,000 years, with phases of weak monsoon rainfall occurring 7500–6800 B.C. and from 3000 B.C. to the beginning of the common era (C.E.) 2,000 years ago. Several of the lakes in Rajasthan dried up completely by 3000 B.C., when, as indicated by a layer of charcoal and ceramics, humans were living in the fossil lake bed.

Palynological evidence from the Rajasthan lake cores largely supports this interpretation (Singh 1971; Singh, Joshi, and Singh 1972; Singh, Chopra, and Singh 1973; Singh et al. 1974; Singh, Wasson, and Agrawal 1990; Prasad and Enzel 2006). For the first half of the Holocene, the landscape was dominated by grasses and herbaceous plants, including an abundance of the shrubby *Artemesia* (a genus that includes Indian mugwort, or *surband*). This genus is associated with a minimum annual rainfall of 500 mm. Wetter climate is also inferred from the presence of *Oldenlandia* (wild chay root, or *daman pappar*), the shallow water grass *Typha angustata*, and the tree species *Syzgium cuminii* (the java plum, or *jamun*), which prefers habitats with a minimum of 850 mm rainfall per annum. These data indicate that mean annual rainfall in Rajasthan prior to 5000 B.C. was as much as 200 mm greater than the current level of 300 mm per annum. A grassy savanna covered this region until about 3000 B.C. (Roberts and Wright 1993), but pollen is largely absent from the soils that accrued after that time (Singh et al. 1974). The absence of pollen in the lake deposits from 3000 B.C. onward indicates that either aridity increased dramatically or conditions for preservation were less favorable.

These sequences developed from the most recent research on the Rajasthan Lake cores radically alter the original climate profile for the Late Holocene proposed by Singh (Singh et al. 1974). That sequence suggested wetter climate 3000–1700 B.C., more arid conditions 1700–1500 B.C., a relatively wet phase 1500–1000 B.C., and then arid conditions returning until 500 B.C. The Singh sequence was one of the first climate sequences that archaeologists could use to understand culture change. Thus it was one of the most influential pieces of research on this topic. It has been broadly applied across South Asia; it was used to suggest an ecological model for the disintegration of the Indus civilization and the decline of the Deccan Chalcolithic settlements. However, more recent research, summarized above, suggests that those early reconstructions were incorrect. The following climate sequence has been inferred for Rajasthan in the latter half of the Holocene: The highest lake levels occurred before 5000 B.C. and a wetter phase persisted after 5000 B.C., lasting until 4200 B.C. The water table was somewhat lower 4200–3500 B.C., and dessication of the lake beds occurred 3500–3000 B.C. The semiarid climate has thus been firmly established in Rajasthan from 3000 B.C. onward.

An examination of carbon isotope ratios from organic deposits in the Lake Lundkaransar cores from Rajasthan further support the early onset of aridity. These demonstrated that a more strongly negative carbon 13:12 ratio occurred before 5000 B.C., and this ratio declined after that time (Prasad and Enzel 2006). Two interpretations were proposed to explain this result. Either there was an abundance of hot-climate-adapted C3 plants surrounding the lake after 5000 B.C. or there was a difference in lake water levels—or both. If lake water levels were lower after 5000 B.C., algae mats may have proliferated. Their photosynthesis activity could deplete the water of dissolved inorganic carbon, causing the less strongly negative isotope ratios. Either way, carbon isotope studies indicate that the arid phase began about 5000 B.C., not 3000 B.C. This reconstruction pushes back the date for the onset of aridity even further, and if correct, it will again dramatically alter our interpretation of the relationship between climate and culture in South Asia. However, these results await independent confirmation. The preponderance of evidence from the lake cores suggests that the semiarid climate trend began at least by 3500 B.C. in Rajasthan.

Paleoclimate Evidence from Other Sources

Analysis of sediment and pollen from cores of the Quaternary sediments underlying the mangrove peat in coastal estuaries of western India demonstrated elevated levels of rainfall occurred along this coastal zone early on in the Holocene, with maximum rainfall occurring approximately 9,000 B.C. and a wet phase that persisted until 2,000 B.C. along the coastline of Maharashtra. This analysis again confirms rainfall levels began to taper off, however, during the third millennium, and the deficiency was sufficient to restrict the growth of mangrove swamps by 2500 B.C. (Kumaran et al. 2005). Deep sea cores from the Arabian Sea also demonstrate another gradual decline in the intensity of the monsoons from 2000 to 1000 B.C. with a significant decline in humidity occurring after 1500 B.C. (Caratini 1991, 1994).

This sequence is further confirmed by palynological studies of lake cores from the Gharwal region of the Himalayas (Phadtare 2000) and varve analysis in Pakistan (Von Rad 1999), which indicated a peak in aridity around 1500 B.C. The varve analysis suggested the arid phase in northwest South Asia was prolonged, reaching a precipitation minimum at 200 B.C. (Von Rad 1999). A somewhat different result was obtained from oxygen isotope ratios in bovid teeth from the Ganga plains. These data confirmed that an arid phase commenced after 3000 B.C. but suggested that the minimum level of rainfall was after 1000 B.C. (Sharma, Joachimski, Sharma et al. 2004; Sharma, Joachimski, Tobschall et al. 2004).

Paleoclimate Data and the Climate-Culture Change Model

The paleoclimate evidence outlined in this chapter converges to strongly challenge the traditional climate-culture change model for explaining the archaeological record in India, as it was originally formulated in the 1970s and 1980s, whereby aridity is associated with the disintegration of settled life in the Indus region after 1900 B.C. and again in the peninsula after 1000 B.C. From all the available evidence it is clear that the monsoon system weakened in South Asia about 5,000 years ago. A semiarid climate phase in South Asia commenced around 3000 B.C. and persisted until the common era in some regions. Thus the mature phase of the Indus civilization (2600–1900 B.C.) occurred within the context of a semiarid climate as did the post-urban phase of the Indus civilization (1900–1000 B.C.). The semiarid climate was well established in peninsular India before 2200 B.C., when Deccan Chalcolithic settlements were springing up in Maharashtra. By the time Chalcolithic people arrived, the western peninsula was already dominated by dry deciduous forest, grassland, thorn scrub, and scrub woodland with a xerophytic, open-grassland vegetation similar to the low-growing, woody, deciduous species growing in this area today (Pappu 1988; Jain 2000). If there was a trend toward greater aridity and a weakening of the monsoon system of a magnitude significant enough to impact human populations, it peaked around 1500 B.C., not 1000 B.C. Monsoon rainfall waxed and waned slightly throughout the second millennium B.C., but Early and Late Jorwe people all lived and died in the context of a semiarid climate.

In the Early Jorwe, Chalcolithic farmers developed coping mechanisms for dealing with the semiarid climate, including strategic land-use practices, with agricultural settlements in Maharashtra located in rocky, hilly regions surrounded by prime agricultural land and satellite herding camps located in large tracts of grassland with shallow soil less suited to farming (Panja 1999). They relied heavily on drought-resistant crops such as barley, maintained diversified planting strategies, utilized crop rotation and double cropping, and maintained foraging, hunting, fishing, and herding activities throughout the second millennium (Mehra 1999). We know that Chalcolithic people employed adaptive diversity, using approximately 80 wild plant species for medicinal purposes, wood carving, building, cooking, and firewood (Kajale 1990; Mehra 1999). Chalcolithic people also may have had less visible strategies for dealing with environmental stressors—behavioral, cultural, and knowledge-based adaptations used to buffer their families and communities from climatic uncertainty during periods of weak monsoon. Rainwater could have been harvested (Pandey, Gupta, and Anderson 2003), irrigation systems could have been devised (Dhavalikar 1988), and plants might have been harvested at different stages in their growth then processed and stored for later consumption (Mehra 1999). Plants also could have served multiple functions, some parts harvested for human subsistence and other parts used for fodder.

Similar to the diverse practices found in modern communities in Maharashtra, Early Jorwe people successfully sustained their communities in the semiarid environment of peninsular India by maintaining their flexibility. Because there is no evidence of a significant large-scale change in climate after 1000 B.C., the archaeological record must provide the clues to decipher what changed in the Late Jorwe phase. If this region had a semiarid climate throughout the second millennium, and Chalcolithic people had developed apparently successful strategies for dealing with their circumstances, then how does the Late Jorwe phase differ and why? What does the archaeological record suggest about why this region was abandoned? How did human populations change in terms of fertility and mortality? Was there a shift in morbidity profiles? Was there any indication that human health changed dramatically? The rest of this book concerns the archaeology and bioarchaeology of the Deccan Chalcolithic and an attempt to answer these questions.

3

◇◇◇◇◇◇◇◇◇◇◇◇◇◇◇◇

Archaeology at Nevasa, Daimabad, and Inamgaon

> Agriculture in India is a gamble with nature; every third year is a bad year
> and every fourth, a famine.
>
> Clutton-Brock (1989)

Deccan Chalcolithic people built mud and mud-brick houses, made wheel-thrown pottery, used copper implements sparingly, and practiced a mixed economy with agricultural, pastoral, hunting, fishing, and gathering elements. The most recent paleoclimate research demonstrates that a semiarid climate phase commenced around 3000 B.C. in South Asia (see chapter 2). Aridity was well established in central India by the time the first villages appeared in the river valleys of Maharashtra after 2200 B.C. and it persisted throughout the Early Jorwe phase as populations grew and the number and size of villages grew. Perhaps to cope with the uncertainty of the semiarid monsoon climate, Jorwe people successfully cobbled together traditions from all over India in a form of adaptive diversity. During this time, local pastoral and gathering traditions mingled with farming practices from northwest and southern India (Dhavalikar 1994b; Shinde 2002; Fuller 2003; Fuller et al. 2004; Fuller 2007). *Hordeum vulgare* (barley) and *Triticum* spp. (wheat) were brought south from Indus Age communities in northwest India and Pakistan. Using *Vigna radiata* (mung beans) and *V. mungo* (black gram) domesticated in southern India, double cropping developed to provide food and fodder for a growing population. Early Jorwe people lived this diverse lifestyle for 400 years until most of the villages in this region were abandoned after 1000 B.C., in the Late Jorwe phase. The archaeological and bioarchaeological record at Nevasa, Daimabad, and Inamgaon provides some clues as to what life was like during this transition and what kinds of changes characterized the Late Jorwe phase.

Nevasa

Nevasa (19°33' N, 74°55' E) is located in the Godavari River valley in northern Maharashtra. The site sits on the southwest side of the Pravara River, a tributary of the Godavari. The remains of the prehistoric community (Fig. 3.1) can be found in three mounds, but excavation was focused on the largest (293 m north to south and 127 m east to west) and deepest deposit (24 m above the current river bed), labeled Mound I (Sankalia, Deo, and Ansari 1960). The majority of the archaeological deposits at the site are undisturbed except for the center of the mound, where a temple was constructed in the first half of the twentieth century, and the edges of the site, where small farms are encroaching.

Eight trenches (5 × 5 m) were excavated at Nevasa down to sterile soil, the layer devoid of cultural deposits. Stone tools close to the base of the mound suggested that the first prehistoric occupation was in the Middle Pleistocene, approximately 150,000 years ago. Hunting and gathering people camped on the mound 25,000 years ago. The first farming community occupied the site in the Early Jorwe phase of the Deccan Chalcolithic, after which a weathered layer of sterile alluvium indicates the site was abandoned[1] until the Early Historic period about 150 B.C. to A.D. 200. The present living village of Nevasa has been occupied since the Maratha period, 600 years ago. It is located east of the archaeological site and extends across to the north bank of the river Pravara.

Although the site of Nevasa is relatively large and well preserved, it is the least understood site of the three included in this book. Excavations occurred here from 1954 to 1956 and again from 1958 to 1959. During this period of Indian archaeology, research was primarily concerned with establishing the culture history and chronology of a site. Archaeologists of that era dug a few very deep trenches and they were focused on documenting and comparing artifact styles across the Deccan region. Theoretical and empirical approaches to archaeology were relatively less sophisticated. Because the excavations were not designed to answer questions about demography, subsistence, or lifestyle, inferences about these topics

Figure 3.1 Aerial view of Nevasa.

are limited, and the following discussion of this site is largely descriptive for this reason.

Excavations at Nevasa revealed no structural features other than floors paved with mud, ash, and cow dung. Lithic artifacts such as hammerstones and anvils, unworked raw materials, unfinished tools in various stages of manufacture, and evidence of tool production (such as debitage) were associated with many of these floors. Consequently, archaeologists identified them as tool-production localities. Chalcolithic people at Nevasa made stone tools similar to those found in other Deccan Chalcolithic sites at Jorwe, Navdatoli, Prakash, and Bahal. Diverse tool types were made from locally available chalcedony, including lunate, triangular, and trapezoidal points, but especially crested-ridge and parallel-sided blades. Novel tool types uncovered at Nevasa included end scrapers on blades, blunted back blades, tools with a tang, borers, drills, and perforators. Polished axes similar to those found at Brahmagiri, Chandoli, Daimabad, and Inamgaon were made from locally available olivine dolorite. All of these tools, including the polished axes, were manufactured in factories at Nevasa.

None of the ceramic vessel sherds found at Nevasa demonstrated any evidence that they were used for cooking food (Sankalia et al. 1960:206). All of the ceramics were utilitarian forms typical of the Jorwe phase, including painted black on red ware, course red and orange ware, which were used for serving, and gray ware, which was used for funerary purposes. Painted black on red ware is the most common type of ceramic at Nevasa in Period III. This is wheel-thrown pottery made from red-bodied clay painted with black slip. Occasionally a red slip was also applied. Forms included a variety of shapes: carinated bowls, spouted vessels, and high-necked pots common to the Jorwe phase. Other forms included spouted bowls, high-necked spouted vessels, basins, and stands. Small bowls were found that had been made using a press mold technique. A few sherds preserved bamboo-mat impressions, possibly suggesting basket molds were used. Most of the ceramics are decorated with "monotonous" designs, though some have carefully executed geometric patterns (Sankalia et al. 1960:206). Paintings depicting animals were discovered on a few sherds, including images of dogs and spiral-horned antelope.

Coarse orange and red ware sherds were ubiquitous in low frequency throughout the Jorwe deposit at Nevasa. These vessels were built from a very coarse clay body and were described as "inferior in firing, in the purity of the clay, in the treatment of the surfaces. The shapes are strictly utilitarian with the predominance of storage jars and dough plates. The former, besides being used for storage purposes, were also used as burial covers for adults" (Sankalia et al. 1960:210). Bright orange ware found in negligible quantity at Nevasa is similar to a ceramic found in the lowest levels at Nasik. Gray ware included three varieties: burnished, coarse, and red-painted, all representing burial urns.

From the archaeological evidence available we can infer subsistence at Nevasa and nearby Daimabad were broadly similar. The presence of groundstone objects such as mullers, querns, and grinding stones indicates that grain and other vegetal foods were processed at the site. A "standardized system of stone weights" may indicate that there was a uniform system for exchanging grain and other produce (Sankalia et al. 1960:476). Little else is known about subsistence at Chalcolithic Nevasa. Appendix 6 of the excavation report contains the extent of information on this subject in a one-page list of seeds and burned grains recovered from the excavation. This list includes cereals, pulses and peas, oil seeds, and spices but it only describes seeds recovered from the Early Historic and later periods. There is no mention of the Chalcolithic period and thus it must be assumed that no such evidence was preserved or collected from Period III at this site. The presence of grinding stones indicates vegetal foods or cereals were obtained through gathering, gardening, agriculture, trade, or some combination of the four.

The inventory of faunal remains is also not temporally subdivided (Sankalia et al. 1960:Appendix 7) and thus provides only general information about the animals in the food refuse. *Bos indicus* (cow) was the domesticated species most prevalent in the assemblage, followed by the wild *Cervus duvaucelli* (barasingha) and *Axis axis* (chital deer). *Capra/ Ovis* (sheep/goat) were uncommon, and *Sus* sp. (pig or wild boar) was represented only by two bones in the Historic period Indo-Roman levels.[2] This inventory of faunal remains at Nevasa was again broadly similar to that at nearby Daimabad. There is thus evidence for cattle keeping, sheep/ goat herding, and hunting. However, further excavation at this site would be necessary to provide clarity about subsistence at Nevasa in the Jorwe phase specifically.

Seventy-one subadult human individuals[3] were buried at Nevasa.[4] The skeletons were sorted and inventoried in 2001 by myself and Veena Mushrif Tripathy. Tripathy devised a new numbering system to account for all of the individuals, and she published morphological descriptions for each individual, including pathological conditions and measurements for each element (Mushrif 2001; Mushrif and Walimbe 2006).[5]

In the Deccan Chalcolithic in general, most of the burials were of subadults, and adult burials are uncommon. At Bhramagiri, Jorwe,[6] Nevasa, Prakashe, Chandoli, Bahal, Khandesh, and Navdatoli,[7] Daimabad, and Inamgaon, subadults were interred under paved floors, inside of large funerary urns, hand built from coarse, gray-bodied ceramic. The urns have a constricted neck, flared mouth, and globular body. Urns of this specific clay body, shape, and type were only used for funerary purposes. The majority of the urns at Nevasa were undecorated, but 6 of the 34 (17%) did have simple decorations, including fingertip impressions at the neck, slanted, short, vertical incisions, rounded chain incisions, a diamond design at the neck, and short incisions on the rim (Sankalia et al. 1960:209).

Double urns were preferred at Nevasa. The term "double urn" refers to a tradition in which two ceramic vessels were placed mouth in mouth, or mouth to mouth, and sealed with clay. One burial (VM 26)[8] contained three urns and only two burials (VM 4 and 16) had a single urn. Two burials (VM 10 and 19) were of older subadults whose bodies were placed in two larger storage jars, not burial urns. The urns were laid directly on the top of the soil, except for the two burials in phase I (VM 22 and 23), which were interred in a shallow pit. The urns were placed in a north-south orientation, except for Burial 9, which was oriented northeast to southwest, perhaps indicating a different season of burial (because of the sun's orientation). Burial VM 14 was laid out directly on a lime plaster floor in a flexed posture, without any burial urns. Burials VM 2 and 28 contained pots of Jorwe ware to cover remains left exposed outside the urn. Burial VM 6 appeared to contain remains from more than one child. Burial VM 16 was described as a symbolic burial and contained no skeletal material.

Grave goods included small painted bowls and small globular vessels (VM 1), high-necked pots (VM 20), faience (VM 7), and carnelian beads (VM 18, 20, 22, 28, and 30). These pieces were constructed of coarse gray ware, which was used for no other ceramic vessels at the site. Of the 594 beads recovered from the Chalcolithic layers (554 steatite, 14 faience, 9 agate, 7 carnelian, 3 shell, 3 terra cotta, 2 amazonite, and 1 each of chalcedony and jasper), only faience, agate, and amazonite were included as grave goods (25, or 4.2% of all the beads recovered). Copper objects were included in a few burials: 2 copper bangles, 7 copper beads, a copper chisel,

1 poker/needle, 1 rod, and 1 copper bowl. Specific details in the excavation report (Sankalia et al. 1960) about ceramics, burial types, and grave goods are limited to descriptions, sometimes without specific burial numbers or any quantitative information. Thus it is difficult to draw specific conclusions from this information.

Daimabad

Daimabad (19°30' N 74°41' E) is located in the Godavari River Valley of northern Maharashtra, approximately 25 km from Nevasa. The archaeological mound sits 120 m north of the Pravara River bed (Fig. 3.2). The western portion of the site was disturbed by construction of a railway line. The river flows on the southern end of the deposit, and cultivated fields have been established on the north side of the mound. The original geographic extent of the deposit is unknown, but at the time of the excavations in 1958–59, the site covered 1,000 × 500 m (Sali 1986). Archaeological deposits are 5 m deep.

Excavations occurred in three phases led by different investigators: 1958–59, 1974–75, and 1975–79. This work demonstrated that Daimabad was occupied for all of the five phases of the Deccan Chalcolithic period. The uppermost 2 m of deposit were from the Early Jorwe phase (1400–1000 B.C.), followed by 80 cm of deposits from the Malwa phase (1600–1400 B.C.), 80 cm of deposits from the Daimabad phase (1800–1600 B.C.), 50 cm of deposits from the Late Harappan phase (2000–1800 B.C.), and 20–30 cm from the Savalda phase (2200–2000 B.C.). The material culture at Daimabad largely resembled other Deccan Chalcolithic sites in ceramic styles and lithic technology, but there are unique artifacts at Daimabad as well, including an agate phallus from the Savalda phase, which may have had religious significance.

Indus styles were more common at Daimabad than other sites from this period. Several items recovered from the Late Harappan phase provide a strong indication of contact between these two cultures—two button-shaped seals with Indus script, four pot sherds inscribed with Indus script, and four large bronze wheeled sculptures. Daimabad is the only Chalcolithic site that yielded bronze objects. The sculptures depict an elephant, buffalo, and rhinoceros standing on wheeled platforms and a man

Figure 3.2. Aerial view of Daimabad.

driving a wheeled chariot. There is one burial of an adult in a brick-lined chamber from the Late Harappan period. The brick sizes conform to Indus standards, which also suggests influence or contact between the two cultures.

Plant remains at Daimabad included charcoal, leaf impressions, seeds, and charred grains (Vishnu-Mittre et al. 1986). Charcoal species indicate the burning of *Acacia* sp. (thorn trees) for fuel throughout the Chalcolithic. Charcoal samples from the Malwa phase were identified as *Anogiessus latifolia* (axelwood tree), *Cassia fistula* (Indian laburnum), and *Dalbergia latifolia* (Indian rosewood).[9] In the Jorwe phase, three different species were used for firewood: *Zyziphus mauritania* (jujube),[10] *Pterocarpus marsupium* (Malabar or Indian kino tree),[11] and *Trema orientalis* (charcoal tree)[12] (Vishnu-Mittre 1981; Vishnu-Mittre et al. 1986). All grow in Maharashtra today in regions that are dominated by dry to moist deciduous forest (Vishnu-Mittre et al. 1986).

Throughout this time, the economy at Daimabad was based on stock raising, hunting, agriculture, and gathering wild foods. Faunal remains indicate Daimabad people kept domesticated *Bos bubulus* (buffalo), *Bos indicus* (cow), *Capra/Ovis* (sheep/goat), and *Canis familiaris* (dog) (Kajale 1977). Wild animal species used for food included *Antelope cervicapra* (blackbuck), *Cervus unicolor* (sambar), *Axis axis* (chital), and *Tetracerus quadricornus* (four-horned antelope) (Kajale 1977). Avian and aquatic species were also exploited for food (Kajale 1977).

Charred seeds and grains indicated that both *kharif* and *rabi* crops were grown. At Daimabad, double cropping began in the Malwa phase and persisted through the Early Jorwe phase (Kajale 1977; Vishnu-Mittre et al. 1986). Charred grains included South Asian and African species brought from northwest India: *Triticum* spp., *Hordeum vulgare*, and *Oryza sativa* (rice). Peas and lentils domesticated in south India were present at Daimabad in the Early Jorwe phase. Pulses present in the floral remains included *Lens esculenta* (lentils), *Pisum arvense* (common pea), *Dolichos biflorus* (horse gram), and *Phaseolus mungo* and *P. aureus* (black and green gram). Seeds used for oil production included *Linum usitatissimum* (linseed) and *Carthamus tinctorius* (safflower). Three additional crop species were used in the Early Jorwe phase: *Eleusine coracana* (finger millet), *Sorghum bicolor* (sorghum millet), and *Paspalum scrobiculatum* (Kodo millet). Daimabad people also used *Zyziphus jujube* (jujube) from the local forests after the Malwa phase.

Double cropping and new cereal crop species introduced in the Early Jorwe phase coincide with increasing population size (Shinde 2002). The settlement covered an increasingly large area through time. In the Savalda phase, Daimabad covered an area of approximately 300 m east-west by 100 m north-south (3 ha or 30,000 m^2). This small village grew in the Late Harappan phase to 20 ha (200,000 m^2). A sterile layer of weathered soil between the Harappan and Daimabad deposits indicated to archaeologists that the site was abandoned in the intervening decades. The site was reoccupied and the settlement remained the same size of 20 ha in the Daimabad and Malwa phases. In the Early Jorwe phase, the site of Daimabad grew to 30 ha.

The number of house floors uncovered in the excavated area (10 × 40 m = 400 m^2 or 4,300 ft^2) grew similarly. There were 11 floors uncovered from layers occupied in the Savalda phase, 15 from the Malwa phase, and 25 from the Early Jorwe phase. The density of the occupation in this excavated area of 400 m^2 changed from 1 household per 36 m^2 (387.5 ft^2) in the Savalda phase to 1 household per 16 m^2 (172.2 ft^2) in the Early Jorwe phase. Archaeologists estimated that 200 people (33 households) occupied each hectare of the village, equivalent to 120 people (20 households) per acre of land. They inferred a total population size of 4,000 residing at Daimabad in the Malwa phase and that the population grew to 6,000 in the Early Jorwe phase (Dhavalikar 1988).

Burial practices at Daimabad were heterogeneous compared to Nevasa and Inamgaon. In the Early Jorwe, infant and child burials occurred in association with structural features. A preference for double urns developed (80% of interments) but burials using funerary jars (10.4%) also occurred, a practice not seen in the Malwa phase. There were more burials with grave goods in the Early Jorwe phase compared to the Malwa phase, and there was a change in the type of grave goods—from a preponderance of beads in the Malwa phase to groundstone in the Jorwe phase.

In total, 71 infants and children were buried in large gray urns and storage jars under house floors in the habitation area at Daimabad (Appendix A). There is evidence that burial traditions changed over time. One adult was buried in a brick-lined tomb constructed in the Late Harappan phase. Two symbolic burials (GR 22 and 45) from the Malwa phase contained ceramics and remains of burned grass—*Poa cynosuroides* (*dharba*), *Desmostachya bipinnata* (*kusha*), or *Agrostis linearis* (*durva*)—but no skeletal remains. One individual buried in the Malwa phase (GR 75) was interred in two remarkable urns decorated with a dog, sun, and peacock motif, and bands were painted on the urn's shoulders. This burial also contained 255 steatite beads, 22 carnelian beads, and 1 ceramic bowl. There was one triple burial (GR 68, 69, and 70) at Daimabad in the Early Jorwe phase. Three individuals were interred together with three groundstone mortars in a double urn buried beneath the floor of House 62. In the Jorwe phase, 45.8 percent of burials at Daimabad contained at least one ceramic vessel. Groundstone mortars were the second most common artifact type in the grave goods (16.7%). Beads were uncommon; six individuals were buried with 11 beads in the Early Jorwe phase; six of these beads were found in a single burial (GR 43).

Inamgaon

Inamgaon (18°36' N, 74°32' E) is located next to the Ghod River, a tributary of the Bhima River (Dhavalikar et al. 1988). The archaeological site covers an area of almost 5 ha (Fig. 3.3). The largest of five archaeological mounds rises 14 m above the river bed. Excavations took place at four of these mounds in 12 seasons from 1969 to 1982, but most of that work was performed at the largest mound (Mound I). At Inamgaon, the horizontal excavation strategy, detailed documentation, collection, analysis, and publication records provide the most thorough reconstruction of life in the Deccan Chalcolithic period. Additionally, 38 radiocarbon dates provide a relatively complete picture of the chronology for each phase of occupation at Inamgaon. From these dates, it is clear that the site was founded in the Malwa phase circa 1565 B.C. and the village grew rapidly to cover 5 ha in the Early Jorwe phase.

Inamgaon also is one of only a handful of sites that preserve a record of the Late Jorwe phase of the Deccan Chalcolithic (Dhavalikar et al. 1988). The following discussion of the archaeological record at Inamgaon is framed as a contrast between these two phases. While the Early Jorwe phase was characterized by population growth and relative prosperity, archaeologists describe village life in the Late Jorwe phase as desperate and poverty-stricken. This interpretation, which I call the climate-culture change model, is based on the following changes noted in the archaeological record: a shift from large, rectangular houses to small, round dwellings; a decline in the amount of agricultural production; a change from cattle keeping to sheep/goat herding and hunting; and "degenerate" material

Figure 3.3. Aerial view of Inamgaon.

culture (ceramics and stone and copper artifacts) in the Late Jorwe phase. In the following paragraphs I describe these aspects of the archaeological record and afterward summarize more recent accepted interpretations.

The excavations at Inamgaon were conducted with a focus on the "house" as the unit of excavation. Houses were identified by the presence of paved floors. In the words of the principle investigator, Dr. Dhavalikar, "I thought the time had come not only for larger, period-wise, excavations, but a house-wise excavation which alone can give insight into the day to day life of the inhabitants. Here every object, small or big, important or insignificant, would be immediately plotted and its full significance in the life of the inhabitants investigated" (Dhavalikar, Sankalia, and Ansari 1988). One hundred thirty-four rectangular and circular floors, some intruded by later pits, were identified in the excavations.

Archaeologists recognized a shift in the occupation at Inamgaon from predominantly rectangular houses constructed in the Early Jorwe phase to round houses constructed in the Late Jorwe phase. In the Early Jorwe phase, 38 percent of the floors at Inamgaon were rectangular. They were large (15 to 35 m²) and positioned in neat rows separated by narrow lanes 1.5 m wide (Dhavalikar 1988). The smaller, circular floors (62%) were haphazardly arranged in small clusters of three to five. In the Late Jorwe phase, archaeologists found that a smaller percentage of the floors were rectangular (50%).

Archaeologists further noted a reduction in the number of floral remains in the Late Jorwe phase, and there is evidence for a simultaneous shift in species preferences away from cereal crops. This evidence has been interpreted as a shift from subsistence based on agriculture to one based on pastoralism, hunting, and gathering. Charred seeds and grains (30,201) were systematically recovered during the excavation and analyzed extensively (Kajale 1988). The number of charred seeds and grains recovered declines from 18,606 in the Early Jorwe phase to 8,782 in the Late Jorwe phase, despite more extensive excavation of Late Jorwe layers. Analysis of these remains demonstrated that despite declines in the volume and number of seeds, six staple crops persisted throughout the sequence (Table 3.1): *Triticum* spp., *Hordeum vulgare*, *Lens esculenta*, *Pisim arvense* (common pea), *Dolichos biflorus*, and *Dolichos lablab* (hyacinth bean). Barley was the most prevalent cereal species in the floral remains; wheat and the common pea were never as common in the floral assemblage as the other species.

Table 3.1. Total number of seeds recovered and proportional representation of species at Inamgaon by layer.

Layer	Total #	Wheat	Barley	Lentils	Pea	Horse gram	Hyacinth	Green pea	Jujube
1	46	0.48	13.04	45.70	0.70	2.17	2.17	0.00	23.91
2	930	1.50	13.44	6.55	0.43	2.50	60.50	0.75	14.30
3	1691	0.71	15.08	35.50	21.90	12.20	8.80	1.06	24.50
4	4070	0.92	29.50	40.61	1.57	6.90	4.20	0.54	15.72
5	2091	3.95	30.20	23.90	5.14	16.24	11.86	1.20	7.53
6	5021	4.12	32.80	20.33	6.73	17.62	13.36	0.48	4.52
7	4337	3.33	31.53	14.88	10.30	22.44	12.42	0.21	4.85
8	3914	1.15	34.50	15.60	1.00	37.00	8.71	0.26	1.71
9	2724	0.55	15.00	10.53	0.33	33.14	39.10	0.11	1.21
10	673	0.44	25.26	15.90	1.04	43.00	10.10	0.74	8.56
11	1937	0.30	26.22	17.03	0.77	47.23	5.93	0.10	2.37
12	771	1.21	24.40	25.03	1.16	40.60	4.30	0.40	2.85
13	1215	0.16	50.40	16.90	0.74	25.40	5.43	0.25	0.74
14	427	0.93	33.25	21.00	1.64	31.85	5.38	3.04	2.81
15	226	0.30	14.54	4.29	1.83	8.27	1.07	3.36	0.92
16	128	0.78	53.90	13.38	0.78	19.53	6.25	1.56	3.90
tl	3021	20.83	443.06	327.13	56.06	366.09	199.58	14.06	120.40

Source: Data from Kajale (1988:727–822).

The proportion of plant species represented also changed through time (Table 3.1). In the Malwa (layers 16–12) and Early Jorwe (layers 11–6) phases, agricultural activities were focused on growing drought-resistant barley during the summer monsoon and growing horse gram in the dry winter months. In the Late Jorwe phase (layers 5–1), farmers planted hyacinth beans during the summer months and lentils during the winter. In the final years of the site's occupation (the uppermost three layers), Late Jorwe people relied less on drought-resistant barley and refocused on wild food resources such as Indian jujube and freshwater mussels in the summer. Indian jujube is a wild fruit that grows in deciduous forest areas and is commonly eaten today, especially in periods of monsoon failure (Kajale 1988). Lentils were still grown, but agriculture played a smaller role over time in the overall subsistence pattern.

Cow bones dominate the faunal assemblage in the Early Jorwe phase (Table 3.2). In the Late Jorwe phase there is an increase in the proportion

Table 3.2. The faunal assemblage from Inamgaon: Minimum number of individuals (%)

Layer	Cattle	Sheep/goat	Antelope[a]	Mollusc	Mammals	Vertebrates[b]	Total
1	215	547	89	195	871	247	1,118
2	541	2,046	155	275	2,777	420	3,197
3	669	2,466	177	238	3,388	417	3,805
4	931	899	304	139	2,173	266	2,439
5	456	257	69	45	808	71	879
6	806	220	84	76	1,126	105	1,231
7	795	267	54	91	1,135	108	1,243
8	1,552	294	16	43	2,047	72	2,119
9	696	121	64	15	913	33	946
10	289	92	42	24	431	36	467
11	658	184	209	27	1,082	66	1,148
12	954	299	217	85	1,493	114	1,607
13	462	144	106	18	708	34	742
14	647	146	136	45	965	78	1,043
15	509	180	154	45	869	63	932
16	212	105	34	25	346	59	405
Total	10,392	8,267	2,062	1,386	21,132	2,189	23,321

Source: Data from Thomas (1988); faunal assemblage: Dhavalikar et al. (1988:823–963).

Notes: a. Antelope includes sambar, chital, and blackbuck.

b. Vertebrates column includes fish, birds, and reptiles

of sheep, goat, and wild foods. Detailed studies were conducted on the 29,473 faunal specimens from Inamgaon (Thomas 1988; Pawankar 1997). Agro-pastoralism was practiced throughout the Chalcolithic period at Inamgaon, but the ratio of faunal remains to soil volume decreases in the Late Jorwe phase. There were 11,438 bones and shells in 28,195 m^3 of excavated soil from the Late Jorwe phase compared to 7,154 bones and shells in 1,611 m^3 of excavated soil from the Early Jorwe phase and 4,729 bones and shells in 978 m^3 of excavated soil from the Malwa phase.

Assuming that there was no change in the disposition of the remains through time, there was a significant decrease in the amount of beef consumed in the Late Jorwe phase. *Bos* sp. comprised the greatest proportion of remains in the Malwa (58.9%) and Early Jorwe (67.0%) phases but made up a small percentage of the remains (24.6%) from the Late Jorwe phase. *Capra/Ovis* increased through time, from 18.5 percent of the remains in the Malwa phase to 16.5 percent of the Early Jorwe phase faunal remains. In the Late Jorwe phase, sheep and goats comprised 54.4 percent of the faunal remains. Although the frequency of the sheep/goat remains changed over time, the use of the animals was more consistent. More than half were slaughtered before 36 months of age in all three phases. In contrast, consumption patterns for cows changed through time as these animals were slaughtered at younger ages in the Malwa and Early Jorwe phases but maintained to older ages in the Late Jorwe phase.

The proportions of antelope taxa are fairly consistent through time. Antelope and deer were the most important wild animal resources in the Malwa phase. They represent 13.7 percent of the faunal assemblage from Malwa layers, versus 6–7 percent of the Early and Late Jorwe assemblages. The majority of the cervid remains were identified as *Antelope cervicapra*. In contemporary India, these animals primarily occupy plains regions and avoid hilly tracts or forested areas. *Cervus unicolor* remains were more common in the Malwa and Early Jorwe phases than in the Late Jorwe phase. This species lives on forested hillsides of Maharashtra. *Axis axis* remains were least abundant throughout the sequence. Chital is the most abundant deer species in India today and typically inhabits grassy forested areas. The least abundant species in the faunal remains is *Tetracerus quadricornus*, a species that is presently abundant in peninsular India and inhabits open forest and tall grasslands. The relatively consistent percentages of antelope in the faunal remains indicates no major changes occurred in the landscape ecology, carrying capacity, or human exploitation of these species over time despite fluctuations in use of other food resources.

Culture Change at Inamgaon: New Interpretations

Climate change has often been invoked as the primary mover of culture change in the second millennium B.C., but the reality of a changing climate around 1000 B.C. has not been substantiated (see chapter 2). Recent paleoclimate research demonstrates that Maharashtra was a semiarid ecozone before the Deccan Chalcolithic began and that these conditions persisted throughout the second millennium B.C. until the common era. Some studies even suggest that the peak of aridity was close to 1500 B.C., just prior to the beginning of the Early Jorwe phase. Certainly annual fluctuations in rainfall also occurred, and these can have important short-term impacts, but the degree of aridity and the magnitude of climate uncertainty do not appear to have increased on a large scale at the beginning of the first millennium B.C. While local environmental changes may have occurred as a consequence of long-term settlement, there is no evidence of significant climate change at the end of the Early Jorwe phase.

New interpretations of the archaeological evidence also have changed the prevailing view of culture change at Inamgaon. Changes in house size and type that were previously interpreted as evidence for culture change through time at Inamgaon can also be seen as evidence for a demographic shift. Although all of the floors were initially identified as houses, more recent research on site formation suggests that such floors may not represent residential structures and that the function of each structure may have changed through time (Panja 1996, 1999, 2003). The sixteen floors uncovered in layers of the Malwa phase have at least one feature (hearth, firepit, and/or storage silo) that suggests a domestic function. As I described above, 38 percent of the floors at Inamgaon were large, rectangular, and neatly positioned in rows in the Early Jorwe, while 62 percent of the floors in this period were smaller, circular, and haphazardly arranged in clusters. However, only the rectangular floors had associated domestic facilities; the circular floors had none. This suggests that the circular floors may have served a different function in the Early Jorwe.

In the Late Jorwe phase, a smaller percentage of the floors were circular (50%) and they were associated with domestic features such as hearths and external courtyards. Rectangular floors built in the Late Jorwe phase had no associated facilities. Thus the meaning of the houses appears to have changed (Panja 1996, 1999, 2003). In the Late Jorwe phase, large areas of the site were uninhabited, and rectangular structures that had been used as houses in the Early Jorwe phase were now used as dump sites for food waste, human bodies, and other refuse. Thus changes in house shape and size do not necessarily reflect a shift in prosperity. The evidence could also be seen as supporting a change in demographic characteristics and a shift in the meaning and function of these floors.

Traditionally, archaeologists have associated round houses in South Asia with poverty *and* greater mobility (Dhavalikar 1994a). It is clear that round houses represent poverty in the contemporary village of Inamgaon. Here people have constructed pit dwellings because "the occupants could not afford wooden posts of adequate lengths and hence in order to obtain the required height for the dwelling, they dug a pit" (Dhavalikar 1994:35). If we accept that the round house has a similar meaning prehistorically, there is no direct evidence that round houses should be associated also with mobility, either presently or throughout prehistory. Similarly, rectangular dwellings are not always associated with settled lifestyle and agriculture (Dhavalikar 1988; Dhavalikar et al. 1988). The association between house shape and mobility or sedentism is so firmly entrenched in South Asian archaeology it has been called the "tyranny of the ethnographic analogy" (Panja 1996). It persists despite conflicting ethnographic evidence. For example, the Kolam is a tribe in contemporary Maharashtra and Andhra Pradesh who build rectangular dwellings in impermanent settlements as part of their lifestyle as shifting cultivators (Dhavalikar 1994:37). House shape should not necessarily be used to infer mobility; however, there is evidence that much of the site was abandoned in the Late Jorwe.

Floral evidence also can be interpreted in ways that do not necessarily support the climate-culture change model. Drought-resistant barley was at its peak in production at the very end of the Early Jorwe phase, when the population size was greatest. In the Late Jorwe phase, people began growing lentils and hyacinth beans instead of barley. These crops continue to be cultivated in Maharashtra today, usually in areas that experience higher rainfall than Inamgaon. Dhavalikar, in support of his climate-culture change model, explained this inconsistency as evidence of an increase in trade between Inamgaon and wetter regions elsewhere in Maharashtra in the Late Jorwe (Dhavalikar 1988). Kajale (1988), on the other hand, suggested that unlike barley, lentils have the advantage of being tolerant of very saline soils. Thus it is likely that high population density in the Early Jorwe phase led to overproduction and over-irrigation of the barley fields, leading to increased soil salinity and, finally, degradation of soil fertility. Early Jorwe farmers may have switched to growing lentils and beans in response to these changes in local soil conditions and the resulting reduction in agricultural yields (Kajale 1988). Lower agricultural yields led to deemphasis on agriculture as a regular subsistence activity. Once barley no longer served as a staple crop, intense double cropping was abandoned and pastoral activities, foraging for wild plants, harvesting mussels, fishing, and hunting became increasingly important in the Late Jorwe phase.

Stock-animal exploitation also changed in the Late Jorwe phase. Cows were an important food item in the Early Jorwe phase, and they were killed for meat at a young age to feed the growing population. Aside from their role as food items, there is evidence for a shift in the function or cultural meaning ascribed to cows over time (Pawankar 1997). Osseous remains indicate cows were still kept in this time and were, in fact, kept to older ages before they were slaughtered (Pawankar and Thomas 1997). Adult cattle are useful not only for meat but also for plowing, milk products, and dung for fuel, manure, plaster, and use in ritual. New meanings for cows are evident in the archaeological record of the Late Jorwe phase, when symbolic objects like terracotta figurines increasingly portrayed humped cows and female deities. The latter were ubiquitous at Inamgaon, but cows were not represented this way in previous phases. Also in the Late Jorwe, limb bones from cow and deer were used to make points, punches, chisels, scrapers, handles, whistles, and other objects with increasing frequency (Dhavalikar and Ansari 1988).

Sheep and goats were more common in the food refuse from the Late Jorwe phase, an emphasis usually interpreted as demonstrating a change from sedentary cattle keeping to mobile pastoralism. The increase in the prevalence of sheep and goat remains coincides with an increasing reliance on lentils as a staple crop, a plant which can provide foddering material for goats in a village setting.

Another detail about the faunal remains that has received less attention is that Late Jorwe people relied more heavily on fish, reptiles, birds, and invertebrates than did earlier inhabitants of the site. Vertebrates (fish, reptiles, and birds) and mollusks comprised 20.2 percent of the Late Jorwe phase faunal assemblage, compared with 9.8 percent of the Early Jorwe phase and 12 percent of the Malwa phase. Most of the eight species of mollusks recovered at Inamgaon are aquatic or terrestrial snails that may be naturally intrusive to the deposits and not food refuse (they comprise about 10% of the total invertebrate assemblage), including aquatic species *Melania striatella tuberculata* and *Digoniostoma pulchella* and terrestrial snails *Cryptozona belangeri*, *Pila globosa*, and *Subulina octona*.

Melania striatella tuberculata (aquatic snail) has a very small body size and prefers organically rich, benthic sediments (Suganan 1995). It is an aquatic gastropod common in South India today. *Digoniostoma pulchella* (aquatic snail) is a small-bodied snail that lives in rivers, lakes, and contemporary paddy fields and can be abundant in the monsoon season (Harinasuta, Bunnag, and Radomyos 1987). *Cryptozona belangeri* and *Pila globosa*[13] are both large terrestrial snails. Pila are adapted to a climate that oscillates between heavy rainfall and periods of drought and are numerous in India and Bangaladesh today. These snails are often harvested in contemporary India and their meat is used to feed prawns. *Subulina octona* is a terrestrial snail, 10–15 mm, commonly found in tropical regions and considered a crop-eating pest today.

Lamellidens marginalis (fresh water mussels) dominate molluscan remains (90% of remains). This species was present throughout all three phases of human occupation, and its prevalence increased through time, from 218 shells in the Malwa phase to 276 in the Early Jorwe phase and 912 in the Late Jorwe phase. Meat from these freshwater mussels was likely consumed and the shells were perforated for ornaments and pendants in the Late Jorwe phase. This species is currently eaten as a starvation food in Bihar and West Bengal (Sugunan 1995). Growing reliance on mussels also indicates either that exploitation of Maharashtra's lakes increased in the Late Jorwe phase or that freshwater mollusks were more often obtained through interregional exchange networks over time. *Xancus pyrum* (chank shell) and *Cypraea arabica* (cowry) were also present at Inamgaon, indicating exchanges with coastal people in the Deccan Chalcolithic period. Chank can be boiled and dried for food (Venkataraman and Chari 1953), and their shells are used for bangles and in the production of lime for plastering house floors (Thomas 1988).

Stone and copper artifacts show aspects of both continuity and change from the Early to Late Jorwe phases. Groundstone saddle querns were used for grinding grain and vegetal foods throughout the Chalcolithic, though they decline in number from 30 in the Early Jorwe to 16 in the Late Jorwe phase. Throughout the Chalcolithic, chalcedony blades were chipped from the nodules gathered from the Ghod River. There are more finished blade tools (48%) in the Late Jorwe phase assemblage than in the Early Jorwe phase assemblage (30%). In contrast, the number of blades taken from each core is much lower, an average of eight blades per core in the Late Jorwe layers versus 32 blades per core in the Early Jorwe layers (Ansari 1988). Round stone balls ("slingballs") and perforated stones are most common in the Late Jorwe phase. These items may have been used for fishing and hunting activities.

Scarce raw materials like copper were used only sparingly throughout the Chalcolithic period. The raw material was gathered locally and smelted at the site. Copper bangles, anklets, beads, and pins were more common in the Early Jorwe phase. Utilitarian chisels, fishhooks, drills, tongs, and crescents were more common in the Late Jorwe phase (Dhavalikar and Ansari 1988). Late Jorwe people appear to have preferred freshwater and

marine shells for ornamentation over copper items (Dhavalikar and Ansari 1988). Although shell beads were not numerous at Inamgaon (180
total), 80 percent of the shell bangles, pendants, and ground shell objects
were recovered from the Late Jorwe layers. Shells made from the intertidal
whelk genus *Oliva* were present in both the Early and Late Jorwe phases.
Beads made from freshwater species and other coastal genera *Conus*, *Ne-
rita*, *Turitella*, and *Cypraea* were primarily restricted to the Late Jorwe
phase. The presence of these marine shells indicates that Late Jorwe people interacted with coastal hunter-gatherers 200 km to the west (Morrison
and Junker 2002; Panja 2003). Deer bone and antler were also made into
decorative beads and pendants during this time. Thus it can be inferred
that there was a shift toward utilization of mollusks for food and decoration in the Late Jorwe. This shift in emphasis undoubtedly also reflects
changes in ideological and cultural features.

Burial traditions also serve as a source of information about culture,
continuity, and change. Out of 255 burials excavated at Inamgaon, 176
were reported in a detailed inventory (Lukacs and Walimbe 1986) and
were subject to a temporal analysis of burial traditions (Raczek 2003).
It is clear that Early and Late Jorwe burial traditions remained remark-
ably consistent for infants and children less than 10 years of age (Raczek
2003). There was a consistent preference for double urn burials through
time. About half of the subadults were interred in twin urns regardless of
specific age. Approximately 40 percent of the burials were non-urn burials in the Early and the Late Jorwe phases, and single urn burials were
uncommon. The proportion of graves that contained grave goods also did
not change significantly; 24 percent of the burials from the Early Jorwe
phase and 28 percent from the Late Jorwe phase had at least one item
interred with the body. One to three small ceramic pieces were usually
associated with the skeleton, either inside or outside the urn. The only
significant difference in the infant urn burials is that Jorwe ware slowly
replaced red/gray ware for use in a funerary context. Red/gray ware was
the preferred material for constructing the funerary vessels in the Malwa
and Early Jorwe phases (50–60%), but this material was less commonly
used in the Late Jorwe phase (28%). This development parallels changes
in ceramic styles for all of the assemblage and is not unique to funerary
vessels.

We can see from the archaeological record that the people at Inamgaon
responded to ecological degradation using the same general strategy of

subsistence diversity they had always relied on to cope with life in the semiarid monsoon climate of peninsular India. Their strategy for coping with local environmental degradation in the Late Jorwe was also to become even more flexible in subsistence activities. In the Late Jorwe phase at Inamgaon, the villagers were able to refocus their subsistence efforts away from farming drought-resistant barley and begin to emphasize foraging and foddering strategies and long-distance trade interactions. These were not new behaviors but represent a shift in the balance of different activities made possible because these subsistence activities, and the knowledge required to pursue them successfully, had been maintained throughout the site's history. These strategies successfully sustained the community for 300 years longer than other villages in the region. One question that remains is why Inamgaon differed from so many other Chalcolithic communities in regard to that flexibility.

The archaeological record indicates that population growth was most rapid at Daimabad as it expanded to a relatively large regional center in the Early Jorwe phase. Pressure from population size may have induced and/or accelerated unsustainable agricultural practices. Sociocultural processes or the development of large institutional frameworks for dealing with life in a large settlement may have prevented flexibility in response to changing circumstances. It is possible that the people of Daimabad simply could not adjust fast enough to maintain a healthy population. In a way, Daimabad may have been *too* successful in the Early Jorwe phase. The settlement was larger, more prosperous, and ultimately less sustainable and less responsive to environmental degradation and declining agricultural production. The archaeological record indicates that for Daimabad, the collapse arrived sooner and hit harder. Small satellite villages such as Nevasa, located on Daimabad's periphery, may have also failed rapidly because the fate of the two communities was too closely intertwined. This strategy of close connections to Daimabad may have served Nevasa in the Early Jorwe phase, or it may have been imposed upon them, or both, but when Daimabad and other regional centers fell, they took the small peripheral settlements like Nevasa with them.

Although the people of Inamgaon were able to persist for 300 years longer than many of the other settlements in the region, eventually this settlement also failed. It is clear that individuals, families, and the village of Inamgaon pursued a response to environmental degradation in the Late Jorwe phase that involved a high degree of flexibility. Different crops were emphasized for agriculture, and hardier species originally domesticated in southern India replaced drought-resistant barley from the northwest. The human relationship to cows, sheep, and goats changed. Villagers interacted with coastal foragers in a relationship that became increasingly important. Locally abundant freshwater mussels, fish, and wild plant foods also became increasingly important. Changes in lifestyle were reflected in changes to material culture. In the Late Jorwe phase, there was a new emphasis on manufacturing stone tools for hunting and on using animal bones to manufacture new types of tools. Terracotta figurines depicting cows were created. Freshwater and marine shells were increasingly used for decoration. There were commensurate declines in formerly important items, which affected the quantity and quality of ceramics and the number of saddle quern grinding stones, copper objects, and other traditional items from the old way of life.

As the settlement density declined in the Late Jorwe phase at Inamgaon, large areas of the site were abandoned and used as dumpsites. Socio-sanitation issues may have become an increasingly powerful factor, causing increases in infant mortality and morbidity through time. The bioarchaeological record provides the best evidence on the topic. Paleodemographic profiles provided in the next chapter will clarify population dynamics that characterized the Early Jorwe phase at Nevasa, Daimabad, and Inamgaon. This analysis will provide information about some of the specific changes that preceded the abandonment of the settlement at Inamgaon. Furthermore, analysis of the pathological profile and biocultural stress levels in the Late Jorwe skeletal series from Inamgaon will enlighten us about conditions in the Late Jorwe and what led to the abandonment of this village.

4

Demography

Fertility, presumed to be unchanging and unfathomable, emerges from these data as the clearest window for viewing the demography of archaeological populations, as well as a surprisingly dynamic component of demographic equations, ancient and modern alike.

McCaa (2002:95)

Bioarchaeologists use age and sex estimates from skeletons to reconstruct demographic profiles that describe the growth rate of a settlement, the sum of births (fertility), deaths (mortality), immigration, and emigration. These demographic profiles are then used to compare relative levels of fertility and mortality among prehistoric populations and to interpret features of the pathological profiles for skeletal populations. Paleodemography represents a difficult challenge, and its techniques have been heavily criticized and reviewed in Bocquet-Appel and Masset (1982), Buikstra, Konigsberg, and Bullington (1986), Hoppa and Vaupel (2002), McCaa (2002), and Bocquet-Appel (2007). In addition to problems with estimating age at death from adult skeletons, other critiques are centered on the assumptions that populations in the past were closed to migration and were stationary (population size was not growing or declining). In response, the field has evolved substantially and innovative statistical approaches to the age pyramid have been developed (Coale and Demeny 1983; Sattenspiel and Harpending 1983; Jackes 1986; Jackes 1992; Konigsberg and Frankenberg 1992; Konigsberg and Frankenberg 1994; Paine and Harpending 1996; Paine 1997; Meindl and Russell 1998; Hoppa and Vaupel 2002).

Recent critiques of paleodemography suggested that perhaps we have been too focused on mortality and have not given fertility enough attention. Intuitively, it would appear that the age structure of a skeletal assemblage should represent the age at which people died and should inform us about mortality hazards and survivorship ratios. In contrast, some paleodemographers have focused recently on examining how fertility shapes the age structure of skeletal populations (Sattenspiel and Harpending 1983; Horowitz, Armelagos, and Wachter 1988; McCaa 1998, 2002). It turns out that fertility has a powerful influence because the force of fertility is concentrated on the moment of birth, whereas the effect of mortality is diffused across the age pyramid because individuals die at all ages (McCaa 1998). Working from this perspective changes the inferences we make from skeletal populations, and in some ways the fertility-centered approach can seem counter-intuitive. For example, using a fertility-centered approach, the number of individuals in each age category reflects the size of the entering cohort. An appropriate illustration of this concept is the baby-boom effect, which influences the shape of the age pyramid for generations. Another example would be that mean age at death, traditionally used as a measure of mortality hazards, instead conveys information about fertility rates—high fertility reduces the mean age at death and low fertility increases the mean age at death.

This chapter provides paleodemographic profiles for three Deccan Chalcolithic villages using a fertility-centered model. This approach (McCaa 1998, 2002) is appropriate for the Deccan Chalcolithic skeletal samples because it accounts specifically for changes in the settlement growth rate, something we know occurred at these sites. To use McCaa's model, an estimate of settlement growth rate is required from the archaeological record (McCaa 1998). This estimate is combined with an estimate of gross reproductive rate (GRR), which is obtained from the skeletons. The two estimates are then used in combination to estimate fertility and life expectancy at birth. GRR is defined as the average number of female offspring born to each woman, assuming (1) she survived to the end of her childbearing years, (2) she conformed to differences in age-specific fertility rates, and (3) there was a 105:100 sex ratio at birth (Last 2001). This statistic is used in demography because it is easier to measure fertility when it is confined to one sex and because it accounts for age-specific effects on fecundity. The total fertility rate (TFR) is estimated from GRR. The TFR, or the total number of live births per woman, is a more general statistic that is relatively meaningless because it does not account for age specific fertility, but it is often used in a comparative framework to examine differences between populations. TFR is calculated very roughly by doubling the GRR.

McCaa (1998, 2002) estimated GRR using proportional hazards ratios of juveniles and adults in the skeletal assemblage (Bocquet-Appel and Masset 1982; Buikstra, Konigsberg, and Bullington 1986). The juvenility index is the proportion of individuals who died between the ages of 5 and 14 years to dead adults >20 years (Bocquet-Appel and Massett 1982), and this method deliberately ignores individuals under 5 years of age. In fact, most methods for paleodemography ignore the youngest infants and children because they "should be" underrepresented in archaeological populations (Angel 1969; Weiss 1973). While this age group is often underrepresented in skeletal populations, particular sites may have a preponderance of infants and children if burial traditions, soil characteristics, and other aspects of preservation lead to preferential preservation of the smallest individuals.

In South Asia, adult burial became uncommon after the mature phase of the Indus Age. Infants and children under five years of age dominate the assemblages from all the peninsular sites in the second millennium B.C. At Inamgaon, Daimabad, Nevasa, Chandoli, and Jorwe (the site for which the Jorwe phase was named), infants and young children were buried inside funerary urns, hand built from a coarse, gray-bodied ceramic with a constricted neck, flared mouth, and globular body (Dhavalikar, Sankalia, and Ansari 1988). Often double urns were used with two vessels placed mouth-in-mouth or mouth-to-mouth and the opening sealed with clay (Raczek 2003). The urns were interred beneath the house floors in what may have been abandoned sections of the habitation area (Panja 1999, 2003). Because the urns were sealed, the bodies were sheltered from sun, wind, water, unfavorable soil pH, and other corrosive elements. These small skeletons were similarly protected from animals, roots, and human activities that destroy fragile subadult skeletons. In addition, recovery of small subadult skeletons was enhanced by the research design used at Inamgaon which focused on the "house" as a unit of excavation.

While the fertility-centered approach to paleodemography is appropriate for these samples, gross reproductive rate could not be derived from the proportional hazards methods recommended for other skeletal series. For this analysis, gross reproductive rate was estimated using a method developed specifically for subadult samples (Robbins, in press). This method is applied to the Deccan Chalcolithic samples to characterize life in the Early Jorwe and to address questions about what specific changes occurred in the Late Jorwe at Inamgaon. Although skeletal samples are not a perfect representation of a living population due to sampling and recovery errors, and paleodemographic profiles cannot represent formerly living populations with high fidelity, these profiles are an important tool in bioarchaeology. They are necessary for comparing relative demographic dynamics among sites but also because demography has an impact on pathological profiles and thus must be considered before bioarchaeologists can compare the health of skeletal samples (Waldron 1987; Wood et al. 1992).

Estimating Demographic Parameters in Subadult Samples

To use a fertility-centered approach to demography, it is necessary to derive an estimate of GRR from the age structure. I developed a regression equation for calculating GRR from subadult samples (Robbins, in press) using reference data from Coale and Demeny (1983) Model West Female life tables, which have been used previously to develop methods for paleodemography (Bocquet-Appel and Masset 1982; Buikstra, Konigsberg, and Bullington 1986; McCaa 1998). I developed the equation from model life tables with mortality levels 1–14 and growth rates between -1 and 2 percent to conform to expectations for archaeological populations. I chose mortality levels 1–14 because these models had fewer individuals dying in the oldest age categories (more than 70 years of age). I chose population growth rates between -1 and 2 percent because they conform to expectations for prehistoric populations (Livi-Bacci 2007). I collected data from 98 tables, including growth rate, crude birth rate, crude death rate, observed GRR, mortality level, and proportions of deaths in each subadult age category (ages 20+ were excluded). I calculated the proportion of deaths as the number of individuals dying within each age category divided by the total number (N) of subadults (less than 20 years of age).

I developed the following quadratic equation to estimate GRR from the proportion of young infants (0–12 months) in a subadult sample:

$$GRR = -2.78 + (7.71 \times {}_{0-1}D_{2-19}) + (34.26 \times {}_{0-1}D_{2-19}{}^2)$$

where ${}_{0-1}D_{2-19}$ is the proportion of infant deaths in the first year of life (44 lunar months–1 year) to other subadult deaths (2–19 years). Figure 4.1 demonstrates that my formula predicts GRR ($F = 81.25$, $P < 0.001$) with a reasonable level of accuracy and precision ($R^2 = 0.9805$, $P < 0.001$). The average difference between the predicted GRR and the observed GRR was

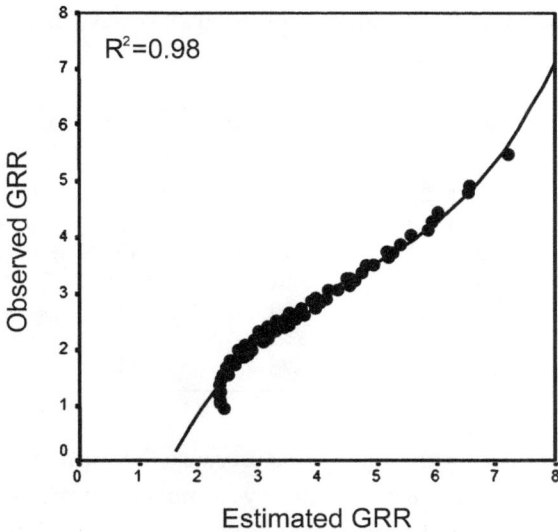

Figure 4.1. Observed versus predicted GRR using estimates of GRR from my formula and Bocquet-Appel's hazard ratio.

1.14 female offspring (Table 4.1). The reference populations had a wide range in the proportion of infants to subadults (from 0.06 to 0.78), and my formula performed with a consistent level of precision across much of this heterogeneous sample. I compared the level of accuracy and precision of estimates from my formula with estimates of GRR from a commonly used proportional hazards method (Bocquet-Appel and Masset 1982).

The estimates for GRR from my formula have a similar degree of accuracy and precision to estimates made using Bocquet-Appel and Masset's ratio (Fig. 4.2). Bocquet-Appel and Masset's ratio performs better for populations with GRR < 2 (total fertility = 4 offspring). Their method tends to underestimate GRR >2. My formula is most accurate for populations with GRR between 2 and 4 (total fertility = 4 to 8 offspring). My formula performs best when the proportion of infants 0–12 months does not exceed 0.45. In populations where more than half of the subadult sample is comprised of individuals 0–12 months (proportion >0.52), the estimate for GRR will exceed 8 (total fertility = 16 offspring). If the proportion of infants 0–12 months in the subadult population is less than 0.13, the estimate for GRR will be fewer than 1.35 female offspring (total fertility = 2.7)

Table 4.1. Sample characteristics and estimates for GRR based on the proportion of infants (0–12 months) in the subadult (< 20 years) sample from 98 Female Model West life tables.

Proportion of perinates	CBR	CDR	Observed GRR	Predicted GRR	SE mean	Obs.—pred.
0.13–0.14	21	24	1.42	2.36	0.005	0.95
0.15–0.19	24	25	1.61	2.45	0.014	0.84
0.20–0.24	30	26	1.94	2.73	0.027	0.79
0.25–0.29	36	28	2.33	3.21	0.040	0.88
0.30–0.34	42	34	2.72	3.81	0.057	1.08
0.35–0.39	49	36	3.25	4.54	0.077	1.30
0.40–0.44	60	45	3.98	5.55	0.123	1.56
0.45–0.49	73	55	5.1	6.76	0.214	1.71

and the margin of error for my formula increases. Because the estimates beyond these parameters would be unreasonable, my formula is appropriate to estimate GRR in samples with proportions of infants (0–12 months) between 0.13 and 0.45 of the subadult population.

Test of the Formula

In his meta-analysis of demographic profiles for 40 historic and prehistoric samples, McCaa (2002) calculated GRR using the Bocquet-Appel ratio. He found that the majority of cemetery samples had a GRR between 2 and 4 (total fertility = 4 to 8 offspring). Contrary to the expectations that infants and perinates "should" be underrepresented in skeletal populations, in McCaa's sample of 40 skeletal populations the mean proportion of subadults to adults was 0.48 (0.22–0.56) and the mean proportion of infants 0–12 years of age to subadults < 20 years was 0.60 (0.24 to 0.84). Because the majority of skeletal populations in that sample had a high proportion of infants to subadults (greater than 0.50), I tested my formula on a subset of these skeletal populations and compared estimates of GRR from Bocquet-Appel's ratio and my formula (Table 4.2) (See Robbins, in press for additional details.) The two methods of predicting GRR produced fairly similar results for skeletal samples with a proportion of infants to subadults less than 0.45. The difference between the two estimates ranged from 0 to 2.7 female offspring. Estimates of GRR differed by more than one female offspring when the proportion of perinates exceeded the upper limit recommended for use of this formula.

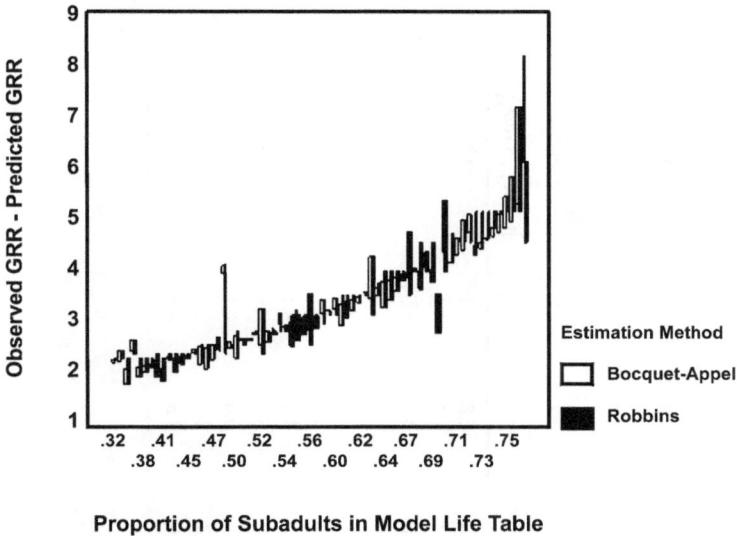

Figure 4.2. Observed versus predicted GRR for 98 model life tables.

I also tested my formula on the skeletal sample from St. Thomas Anglican Church's nineteenth-century skeletal sample in Belleville, Ontario. This large osteological collection is comprised of a high proportion of subadults, similar to the Deccan Chalcolithic samples, and it is a good population for testing demography methods because it is derived from a large cemetery with 1,564 individuals buried between 1821 and 1874 with detailed historical records available for comparison (Saunders, Herring, and Boyce 1995). The sample I used consisted of 575 individuals excavated from 579 grave shafts prior to the construction of a parish hall in 1989 (Saunders 2002).

In the Belleville skeletal population,[1] the proportion of infants 0–12 months to subadults is 0.305. My formula predicts GRR = 3.63 (total fertility = 7.26), a value consistent with the GRR estimates from the Bocquet-Appel ratio (McCaa 2002). McCaa predicted that the best-fitting model for GRR was 3.5 (range = 3.0 to 3.5) if life expectancy at birth in this population was 20 years. The estimate for GRR is lower when the estimate for life expectancy is higher. My formula predicts GRR was 0.13 female offspring higher than the GRR estimate from McCaa's best-fitting model (p >0.95) (McCaa 2002). The two predictions for GRR at Belleville are not significantly different.

Table 4.2. Test of Robbins formula on 11 prehistoric skeletal samples in comparison with estimates for GRR using Bocquet-Appel's ratio.

Population	Infants 0–1 yr	Subadults 2–19 yrs	Proportion $_{0-1}D_{2-19}$	GRR$_{est1}$ $_{5-14}D_{20+}$			GRR$_{est2}$ $_{0-1}D_{2-19}$	Precision $\lvert$GRR$_{est1}$-GRR$_{est2}\rvert$
				$e_x=20$	$e_x=30$	$e_x=40$		
Amelia Island	20	83	0.24	3.1	2.8	2.8	1.1	1.7
Dickson Mound	25	56	0.45	5	4.4	4.3	6.2	1.2
Chiribaya	39	152	0.26	3.2	3.2	3.1	1.54	1.6
Estuquina	107	214	0.50	6.2	5.4	5.2	7.5	1.3
Hawikku	40	83	0.48	4.3	3.8	3.7	7.0	2.7
Loisy-en-Brie	19	50	0.38	3.6	3.2	3.2	4.8	1.2
Maitas	21	55	0.38	4.8	4.2	4.1	4.8	0.0
Monongahela	31	60	0.52	6	5.2	5	7.9	1.9
Pearson	23	52	0.44	.	.	6.6	6.1	0.5
Scarborough	9	37	0.24	3.6	3.2	3.1	2.9	0.2
Tlatilco 4	12	34	0.35	2.8	2.5	2.5	4.3	1.5

Source: Data on age structure from McCaa (1998).

Based on these tests of my formula, it provides accurate and precise estimates for GRR relative to other methods in standard use for paleodemography. Conservatively, the formula is applicable to skeletal samples with a proportion of young infants (0–12 months) to subadults between 0.13 and 0.45. The proportion of infants (0–12 months) in the Deccan Chalcolithic samples ranged from 0.27 to 0.36; thus, I used my formula to estimate GRR for each of the Chalcolithic samples. The estimates for GRR were combined with archaeological estimates for settlement growth rate to derive the crude birth rate and life expectancy at birth, following McCaa (2002:100).

Age Estimation

A full inventory of the remains from Nevasa, Daimabad, and Inamgaon is available and readers are referred to those publications for details on preservation and inventories of the remains (Sankalia, Deo, and Ansari 1960; Kennedy and Malhotra 1966; Lukacs and Walimbe 1986; Walimbe 1986; Mushrif and Walimbe 2006). Age was estimated for skeletons from Inamgaon (Lukacs and Walimbe 1986), Nevasa (Kennedy and Malhotra 1966), and Daimabad (Walimbe 1986). Because methods for age estimation are revised over time and new methods are developed, I assessed age at death independently for this study (Appendix B). Sex estimation was not attempted because the majority of individuals were less than five years of age and had not yet developed mature secondary sex characteristics.

I used long-bone lengths and dental eruption as a technique for age estimation. Age estimates for the smallest, perinatal skeletons (individuals who died prior to the age of one month postnatal life) are derived from measurements of the basilar occipital and lateralis occipital, length of the petrous portion of the temporal or greater wing of the sphenoid, length of the malars, and length of the clavicle, ilium, ischium, and pubis (Fazekas and Kósa 1978; Scheuer and Black 2000). I also estimated age for each individual using regression formulas for long-bone lengths for the humerus, radius, ulna, femur, and tibia (left side if available) (Scheuer, Musgrave, and Evans 1980; Sherwood et al. 2000) and used these estimates to calculate the average age for each individual (Table 4.3). The average difference in the age estimates for a single individual between different bones and different formulas was 0.32 lunar weeks (gestational weeks).

When age is predicted from long-bone lengths using regression, the estimates can suffer from a centrist tendency, by which the age of younger individuals is overestimated and the age of older individuals is underestimated. Age estimation methods developed from regression can also cause the target population to "mirror" the reference population (Bocquet-Appel and Masset 1982; Konigsberg and Frankenberg 1992; Konigsberg and Frankenberg 1994; Bocquet-Appel and Masset 1996; Lucy et al. 1996). The regression equations for estimating age in perinates were developed using bone lengths measured from ultrasound or x-ray films, so this technique also introduces some additional error into the age estimates when these formulas are then applied to dry bone lengths.

Paleodemography requires accurate age estimates. It is particularly important to have confidence in the assignment of individuals to the perinatal and infant age categories because the demographic profile for these samples will be based largely on the proportion of these individuals. To avoid some of these problems, I used a more statistically powerful Maximum Likelihood approach to age estimation from long-bone length (Gowland and Chamberlain 2002; Tocheri et al. 2005). Prior probabilities for age given long-bone length were selected from a large sample of 20,000 individuals (Butler and Alberman 1969). I applied the following formula to the Chalcolithic sample to derive the perinatal age pyramid:

$$p(A_i|L_j) = \frac{p(L_j|A_i)}{\Sigma\,[p(L_j|A_i) \times p(A_i)]} \times p(A_i)$$

where $p(A_i|L_j)$ is the probability of being in age category i given long-bone length j, which is derived from $p(L_j|A_i)$ the probability of having long-bone length j given age category i, $\Sigma\,[p(L_j|A_i)$ is the sum of probabilities of a particular length given age across all age categories, and $p(A_i)$ is the prior probability obtained from the life table for the age category given a certain length.

A Kolmogorov-Smirnov test for two independent samples was used to determine whether the age distributions for the archaeological samples were significantly different from the Perinatal Fertility Survey (PFS) reference sample (Table 4.4). The PFS population is used here to represent the range of variation in human infant sizes at birth. The null hypothesis is that the archaeological samples will not differ significantly from the PFS population. Results of this analysis suggest that there is no significant

Table 4.3. Age estimates from perinatal long-bone lengths (in lunar weeks).

EARLY JORWE INAMGAON

Individuals	n[a]	Scheuer[b]	Sherwood[c]	Mean age
47	3	31.5	30.9	31.2
103a	3	36.6	37.4	37.0
74	1	36.7	37.5	37.1
97	4	37.2	37.9	37.5
78	1	38.7	39.4	39.1
124	1	39.3	40.2	39.7
80	1	40.0	40.9	40.5
103b	1	39.3	41.8	40.5
84	2	40.0	41.7	40.8
83	1	40.8	43.1	42.0
Tl = 10	Tl = 18			

LATE JORWE INAMGAON

142	5	29.9	29.8	29.8
215	1	36.7	37.8	37.3
144	4	38.0	38.6	38.3
18	1	38.1	38.9	38.5
1	4	39.1	40.4	39.7
170	4	39.1	40.3	39.7
118	4	39.4	40.7	40.0
69	1	39.8	41.7	40.8
239	1	40.4	41.7	41.0
63b	1	40.8	42.1	41.4
155	1	40.8	43.1	41.9
63a	3	42.4	43.2	42.8
Tl = 12	Tl = 31			

NEVASA

7	1	40.8	43.2	42.0
17	1	37.2	39.5	38.4
23	4	38.3	39.2	38.8
6	1	39.2	40.2	39.7
Tl = 4	Tl = 7			

DAIMABAD

30	1	35.9	36.7	36.3
34	2	36.4	37.1	36.8
28	4	36.7	37.4	37.1
14	3	37.4	38.0	37.7
11	1	36.7	38.9	37.8
33	2	37.4	38.3	37.8
10	2	38.9	39.8	39.4
35	1	38.7	40.3	39.5
15	5	39.4	40.4	39.9
16	1	39.7	41.9	40.8
Tl = 10	Tl = 22			

Notes: a. n = number of long bones available; e.g., n = 5 means that one humerus, radius, ulna, femur, and tibia was available for that individual.
b. Average age estimate for all bones using regression equations from Scheuer and colleagues (1980).
c. Average age estimate for all bones using regression equations from Sherwood et al. (2000).

Table 4.4. Proportion of perinatal individuals in each gestational age category.

Age (weeks)	PFS[a]	Inamgaon EJ	Inamgaon LJ	Nevasa	Daimabad
16	0.02				
18	0.04	.	.	.	.
20	0.03	.	.	.	.
22	0.03	.	.	.	.
24	0.03	.	0.02	.	.
26	0.04	0.01	0.03	.	.
28	0.03	.	0.01	.	.
30	0.03	0.04	.	.	.
32	0.02	0.04	.	.	0.02
34	0.02	0.01	.	.	0.02
36	0.02	0.07	0.05	0.05	0.16
38	0.03	0.15	0.16	0.17	0.28
40	0.11	0.15	0.25	0.32	0.23
42	0.11	0.15	0.21	0.24	0.18
44	0.11	0.15	0.18	0.18	0.10
46	0.21	0.13	0.05	0.02	0.01
48	0.09	0.09	0.03	0.02	0.01
Z^b		1.069	0.802	0.802	0.802
p-value		0.203	0.541	0.541	0.541

Notes: a. Perinatal Fertility Survey population (Butler and Alberman 1969). Prior probabilities for age estimation were developed from this population following Tocheri et al. (2005).
b. Kolmogorov-Smirnov test for two independent samples was used to test for significant differences between the arcaheological samples and the Perinatal Fertility Survey population.

difference detectable between the age structure of the Deccan Chalcolithic skeletal samples and a sample of 20,000 births (Fig. 4.1). A difference would be expected if there was a significant bias in the preservation, recovery, or representation of perinatal skeletons. A difference would also be expected if the Deccan Chalcolithic populations differed dramatically from the contemporary reference population in terms of infant length at term (i.e., infants on average were smaller) or if one of these populations departed significantly in maternal/fetal health status. I used the results of this analysis to create a category of "perinates" (individuals >44 lunar weeks to 1 year) for the demographic analysis. Results of this analysis suggest that there is no significant difference detectable between the age structure of the Deccan Chalcolithic skeletal samples and a sample of 20,000 births (Fig. 4.3).

For older infants and children who died between the ages of 1 and 60 months, I preferentially used age estimates based on dental developmental timing and eruption (Moorrees, Fanning, and Hunt 1963) when teeth

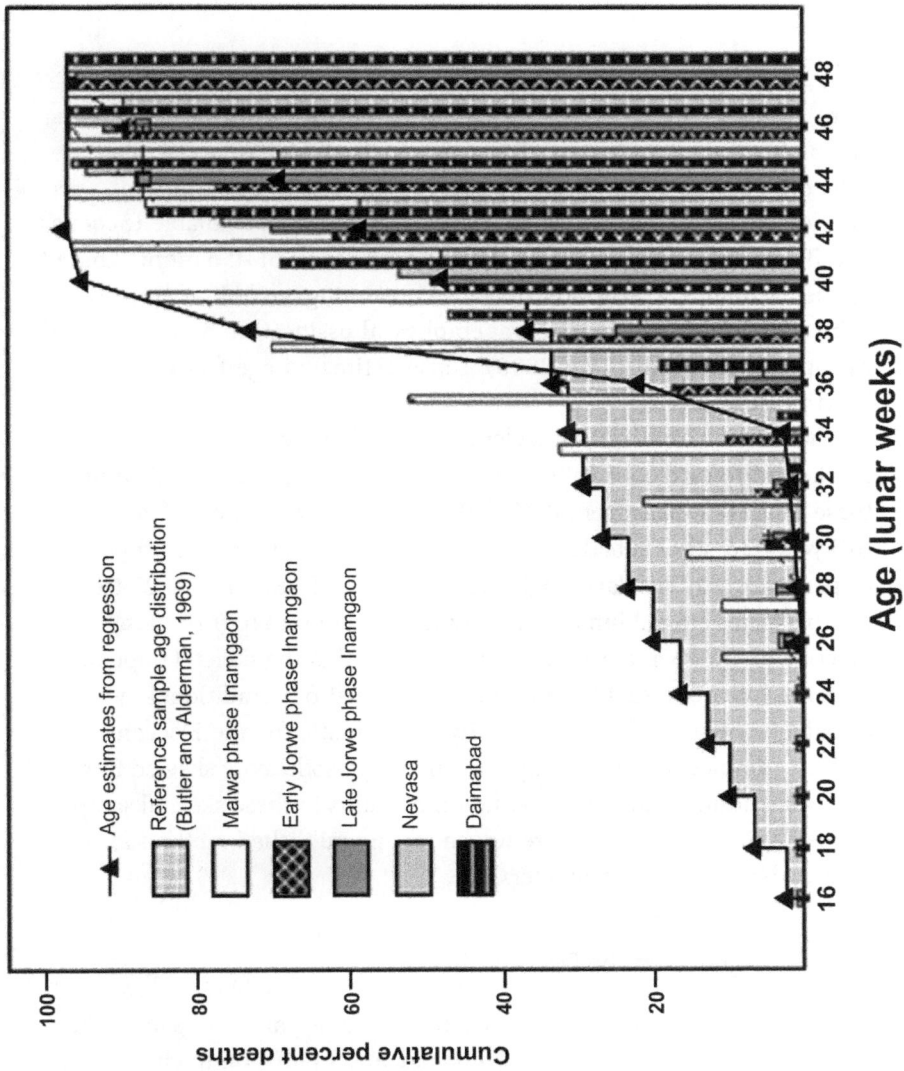

Figure 4.3. Cumulative frequency of perinatal age distribution using prior probabilities.

Table 4.5. Subadult versus adult deaths.

| | Inamgaon | | | | Nevasa | | Daimabad | |
| | EJ | | LJ | | | | | |
	n	%	n	%	n	%	n	%
Adults	4	0.08	24	0.14	3	0.04	1	0.03
n Subadults	44	0.92	122	0.86	71	0.96	37	0.97
N	48	146	74	38				

were available. I also estimated age based on fusion of the mandibular symphysis and metopic suture, measurements of the basilar occipital lateralis occipital, and length of the petrous portion of the temporals, pelvis elements, and clavicle when those elements were available (Scheuer and Black 2000). Age was estimated for individuals 61–156 months based primarily on dental eruption (Moorrees, Fanning, and Hunt 1963; Smith 1991) and long-bone or vertebral epiphyseal ossification and epiphyseal fusion (Scheuer and Black 2000). I also estimated age from long-bone length (Maresh 1970).

The Deccan Chalcolithic skeletal samples have an age structure heavily skewed toward subadults (Table 4.5). The proportion of subadults in these samples ranges from 0.86 to 0.97. The proportion of infants (birth to 3 years) to other subadults was between 0.56 and 0.73. The proportion of perinates was similarly high, with infants 0–1 year from 0.27 to 0.36 and perinates (16–42 lunar weeks) ranging from 0.13 to 0.24 (Table 4.6). Paleodemography is always a challenge given that the skeletal population does not fully represent the entire living population. Traditional methods of paleodemography are particularly problematic for the Deccan Chalcolithic samples because the age pyramids are so heavily skewed toward young subadults. Underrepresentation of adults in these skeletal samples created a situation where there was in fact no published method for estimating demographic parameters.

Paleodemography in the Deccan Chalcolithic

My goal in this chapter is to estimate demographic parameters for archaeological samples from Nevasa, Daimabad, and Inamgaon to characterize the heterogeneity present in the Early Jorwe sites along an urban to rural

Table 4.6. Subadult deaths by age category.

| | Inamgaon | | | | | | | |
| | EJ | | LJ | | Nevasa | | Daimabad[a] | |
Age category (months)	n	%	n	%	n	%	n	%
≤42 lunar weeks	9	.20	18	.15	9	.13	13	.35
1 (1–12)	16	.36	40	.33	21	.30	10	.27
2 (13–24)	4	.09	12	.10	22	.31	8	.22
3 (25–36)	4	.09	10	.08	8	.11		
4 (37–48)	5	.11	5	.04	4	.06	4	.11
5 (49–60)	1	.02	5	.04	1	.01	1	.03
6 (61–72)	1	.02	6	.05	1	.01		
7 (73–84)	1	.02	2	.02				
8 (85–96)	1	.02	5	.04			1	.03
9 (97–108)	1	.02						
10 (109–120)			2	.02	2	.03		
11 (121–132)			3	.02				
12–14 yr	1	.02	6	.05				
14–16 yr			4	.03				
Uncertain			4	.03	3	.04		

Notes: a. one individual included in Walimbe's report of the remains from Daimabad could not be located (may be located at Archaeological Survey of India, Nagpur).

continuum and to understand what changed in the Early to Late Jorwe transition at Inamgaon. The demographic profile can provide insights into the difference between a large, fast-growing regional center like Daimabad as opposed to the smaller, more peripheral satellite settlement at Nevasa. Given archaeological evidence that indicates the Early Jorwe phase at Inamgaon represented a growing settlement and the Late Jorwe phase represented a declining settlement, the paleodemographic profile will demonstrate whether the settlement's decline was due to changes in fertility or life expectancy at birth, or both. If the demographic profile for the Late Jorwe phase demonstrates a high-pressure situation, with relatively high fertility and high infant mortality accompanying declines in settlement growth rate, this result will be seen as support for the portion of Dhavalikar's model that predicts a harsher set of circumstances just prior to collapse at Inamgaon. If, on the other hand, it appears that the Late Jorwe was a low-pressure demographic situation, with declines in fertility and moderate mortality rates leading to depopulation, this result

Table 4.7. Estimates for GRR, crude birth rate, and life expectancy at birth (in years) for Deccan Chalcolithic samples.

Site	$_{0-1}D_{2-19}$	GRR	CBR	-2%	-1%	0%	1%	2%
Nevasa	0.30	2.6	40	14	19	**25**	33	45
Daimabad	0.27	1.8	20	26	35	50	**62**	-
EJ Inamgaon	0.36	4.4	51	12	15	20	**25**	33
LJ Inamgaon	0.33	3.5	65	10	**12**	15	19	24

would support the part of Lukacs and Walimbe's model that predicts an improvement in circumstances in the Late Jorwe.

Archaeological estimates for the settlement growth rate were used in combination with estimates for GRR from the proportion of infants 0–1 year in the subadult skeletal samples to estimate crude birth rate and life expectancy at birth (following McCaa 1998, 2002). Nevasa was a small (2-ha) site close to Daimabad. This peripheral settlement was only occupied in the Early Jorwe phase of the Chalcolithic. Throughout this time, the size of the settlement and the number of floors remained constant. Archaeologists have thereby characterized this village as having a small, stable population size of approximately 400 people (Dhavalikar 1988, 1997). The proportion of infants 0–1 year old in the skeletal population from Nevasa was 0.30, which yielded an estimate for GRR of 2.6. Given a stable settlement growth rate at Nevasa (near zero) and an estimated GRR of 2.6 (TFR = 5.2), the crude birth rate was 40 per 1,000 people and life expectancy at birth was 25 years (Table 4.7). This settlement thus had moderate fertility and a short life expectancy at birth (Fig. 4.4).

Archaeologists suggested the site of Daimabad was growing rapidly in the Early Jorwe phase, from 20 to 30 hectares. Population size increased from approximately 4,000 people in the Malwa phase to approximately 6,000 people 200 years later in the Early Jorwe phase. The proportion of infants 0–1 year of age to total subadults was 0.27, which yielded an estimate for GRR using my formula of 1.8 female offspring, which indicates a TFR of 3.6 offspring per female in this sample. Given this estimate for GRR and a settlement growth rate of 1 percent (indicating that the population was doubling roughly every second generation), the crude birth rate for Daimabad was 20 births per 1,000 people and life expectancy at birth would have been between 60–65 years. This profile suggests relatively good reproductive success, with population growth accomplished through relatively low fertility and high life expectancy at birth.

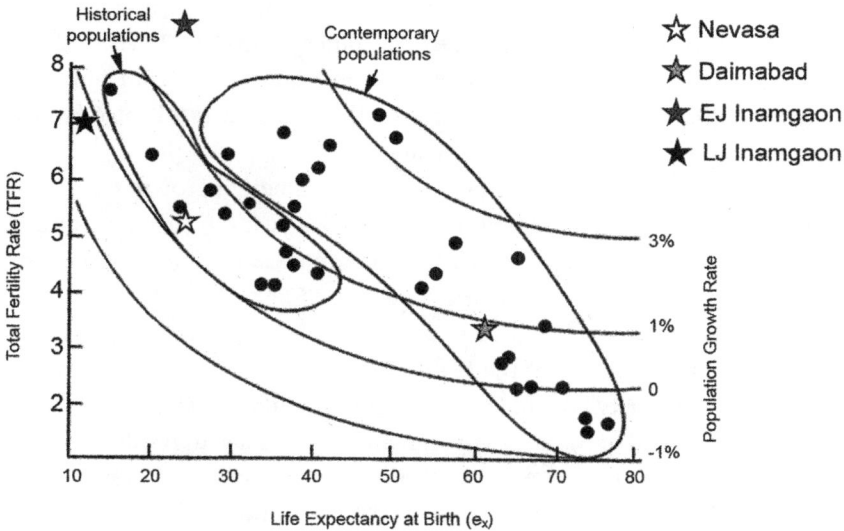

Figure 4.4. Average number of offspring per woman (TFR) versus life expectancy at birth (e_0) in Deccan Chalcolithic samples.

The estimate for life expectancy at birth at Daimabad may be inflated by sampling or methodological error, or because the demographic profile does not account for immigration. Roughly half the skeletons excavated from Daimabad were available for this analysis, potentially leading to sampling error (see chapter 3 for details), and Daimabad had the highest proportion of perinates < 44 lunar weeks in the assemblage (0.35). If the proportion of the youngest infants was different in the larger sample, or if some of these very small perinates were carried to term, this would change the estimate for GRR. Because settlement growth rate is the sum of births, deaths, and migration, rapid settlement growth rate at Daimabad may also have been due to immigration. If we use the demographic profile for Daimabad to indicate *relative* fertility and mortality rates, it is possible to infer that this site had high settlement growth rates, relatively low fertility rates, and a long life expectancy, and that some of the settlement growth may have come from immigration.

Early Jorwe Inamgaon was a regional center of moderate size (5 ha), which grew in the Early Jorwe phase to an estimated population size of 1,000. Based on the proportion of infants 0–1 year of 0.36, my estimate for GRR was 4.4 (TFR = 8.8). Based on this estimate and a moderate growth rate of 1 percent, Early Jorwe Inamgaon would have had a crude birth rate of 51 births per 1,000 and a life expectancy at birth of 25 years. These results indicate that this site was relatively less successful than Daimabad, experiencing moderate settlement growth rates as a result of relatively higher fertility and lower life expectancy.

In the Late Jorwe phase at Inamgaon, the settlement growth rate was in decline. Based on the proportion of infants 0–1 year old (0.33), I estimate that GRR was 3.5 (TFR = 7.0). If we use a modest decline in settlement growth rate (1%), the crude birth rate was 65 births per 1,000 and the life expectancy at birth was 12 years. When life expectancy is less than 20 years, three-quarters of the females born will die before reaching reproductive maturity. To maintain a stable population size under those circumstances, all of the surviving females would have to produce at least eight offspring that survived to adulthood (Johansson and Owsley 2002). Historically, chronic warfare, epidemic disease, and modern urban poverty have temporarily reduced life expectancy at birth to as low as 10 years. It would not be possible to actually have a TFR this high under these circumstances. Because paleodemographic estimates cannot account for migration, and because the life expectancy at birth was lower than a sustainable level, it is likely that the actual life expectancy at birth was higher than 12 years and part of the negative growth rate can be explained by an increase in emigration away from the site. Clearly, the negative growth rate combined with relatively high fertility and relatively low life expectancy at birth indicates that the settlement faced a relatively grave, high-pressure demographic situation.

Biodemography and Culture Change in the Late Jorwe Phase

It is difficult to make inferences about paleodemography in any archaeological sample, but this is particularly true when the assemblage is primarily comprised of subadult individuals under the age of five years. The age structure makes traditional calculations of fertility impossible using methods developed by Bocquet-Appel and Masset (D_{5-14}/D_{20+}) or Buikstra and colleagues (D_5/D_{30}), for example. To apply the fertility-centered model of demography (McCaa 1998, 2002) to these skeletal samples, it was necessary to devise a new way of predicting GRR (Robbins, in press). The formula applied in this chapter was developed from 98 model life tables (Coale and Demeny 1983) and the proportion of perinatal individuals in a sample has a strong correlation with the GRR for the Female Model West life tables (R^2 = 0.98). The margin of error is +/- 1.05 female offspring. Estimates of GRR from my formula compare favorably to the accuracy and precision of other methods in cases where the proportion of infants 0–12 months to subadults (< 20 years) is between 0.13 and 0.45. GRR estimates are less accurate for populations with a GRR less than 2 or greater than 6. If the paleodemographic profiles for the Deccan Chalcolithic are not entirely accurate, they at least provide the first description of fertility and mortality for these samples that can serve as a baseline for comparing these samples in bioarchaeological research.

At the beginning of this section, I posed several research questions for the demographic analysis of the Deccan Chalcolithic samples concerning relative rates of fertility, mortality, and life expectancy at birth at Nevasa, Daimabad, and Inamgaon. I asked how these rates related to settlement size in the Early Jorwe phase and what they indicated about culture change in the Late Jorwe phase at Inamgaon. The pattern that emerges from the demographic profiles for the Early Jorwe phase of the Deccan Chalcolithic is that the small peripheral settlement of Nevasa and the large regional center at Daimabad had significantly different profiles. Daimabad was a regional center in the Godavari River valley, and the settlement was experiencing a positive growth rate in the Deccan Chalcolithic period prior to 1000 B.C. The population accomplished this growth with a relatively low crude birth rate of 20/1,000 and relatively long life expectancy at birth of 62 years. The small satellite settlement of Nevasa had a stable population growth rate throughout the Chalcolithic period, high crude birth rate of 40/1,000, and lower life expectancy at birth of 25 years. Daimabad accomplished faster settlement growth through moderate to low fertility, long life expectancy, and, probably, a contribution from immigration. In contrast, Nevasa was not so fortunate. That population suffered from a relatively low life expectancy at birth and relatively higher fertility rates. Emigration may have also played a role in shaping this profile.

The demographic profiles for Nevasa and Daimabad conform to general expectations for historical populations (Livi-Bacci 2007). Similarly, the regional center at Daimabad conforms to expectations for historical and contemporary populations experiencing relatively fast growth rates (Figure 4.4). The GRR for archaeological and historic cemetery samples is expected to be between 2 and 4.5. The average TFR for historical and contemporary samples lies between 5 and 9 children per female. Life expectancy at birth for human populations in prehistory was generally in the range of 20–40 years and the majority of prehistoric communities probably maintained a slow rate of change that fluctuated between -1 and 2 percent (Livi-Bacci 2007).

Early Jorwe Inamgaon achieved a fast settlement growth comparable to Daimabad but with relatively higher reproductive costs for women. Despite a low life expectancy at birth of 25 years for the Early Jorwe Inamgaon sample, the population was experiencing rapid growth. This was accomplished through higher fertility, with a crude birth rate of 51/1,000. This site is an outlier among Early Jorwe settlements in regard to fertility rates. Archaeologists agree that the Late Jorwe phase at Inamgaon was experiencing a declining population size (Dhavalikar 1988; Panja 1996, 1999). The demographic profile for the Late Jorwe phase indicates that fertility rates were highest (65/1,000) and these were accompanied by a drastic reduction in life expectancy at birth (12 years). Because the life expectancy at birth is at a level that is unsustainable for 300 years of Late Jorwe phase occupation, the negative settlement growth rate was apparently due to combination of relatively high fertility, low life expectancy, high infant mortality, and increased levels of emigration.

Significant deviations from expectations for human populations must be examined critically for evidence of migration, sociocultural and ecological variables that may play a role in shaping the demographic profiles. The pattern at Inamgaon supports the part of Dhavalikar's model that suggested the Late Jorwe represented declining circumstances with dire consequences for human populations. The transition from the Early to Late Jorwe phase at Inamgaon was accompanied by increasingly stressful demographic dynamics that would effectively indicate a demographic collapse occurred in the Late Jorwe phase. This result also supports a suggestion by Panja (Panja 1996, 1999, 2003) that emigration away from the site in the Late Jorwe phase led to abandonment of most of the village at Inamgaon. Panja argued that the function of the houses in the Late Jorwe shifted from residential use to burial sites and garbage dumps.

Results of this analysis do not indicate support for the subsistence transition model proposed by Lukacs and Walimbe. That model suggests that the Late Jorwe was a period of lower stress levels. The demographic profile also indicates that the improvements in health status in the Late Jorwe phase, which were inferred from the lower rate of LHPC, may have been affected by changing demographic circumstances. Rates of LHPC decline through time at Inamgaon and life expectancy at birth demonstrates a similar decline. One explanation is that the demographic differences are driving the differences in the pathological profile. This hypothesis will be subject to further scrutiny in chapter 6, when I compare population differences in subadult health using biocultural stress markers.

5

◇◇◇◇◇◇◇◇◇◇◇◇◇◇

Estimating Body Mass
in the Subadult Skeleton

Bioarchaeologists primarily focus on understanding the lifestyles of past people using morphological variation, markers of diet, pathology, and growth disruption in human skeletons. In this chapter, I demonstrate a method for estimating body mass from the subadult skeleton using a measure of midshaft cross-section geometry (J) (Robbins, Sciulli, and Blatt 2010). In chapter 6, I will use this technique to estimate body mass in the subadult skeletons from Nevasa, Daiambad, and Inamgaon and use these estimates of body mass scaled to stature squared as a biocultural stress marker to test hypotheses about biodemography and stress levels in the Early and the Late Jorwe phases of the Deccan Chalcolithic.

To make accurate predictions about prehistoric populations, bioarchaeologists need methods for estimating age, sex, stature, and body mass (weight) using human skeletons. Body mass can be estimated from the limb bones of the skeleton (also known as long bones) because bone is a dynamic tissue, growing and changing shape throughout the lifespan in response to genetic and ontogenetic processes. Long bones have structural support, movement, and mineral storage functions that determine, and are determined by, such features as bone length, articular dimensions, midshaft cross-section geometry, tissue composition and architecture, mineral density, and porosity (Ruff 1995; Cowin 2001; Lanyon and Skerry 2001; Auerbach and Ruff 2004; Pearson and Lieberman 2004; Ruff, Holt, and Trinkaus 2006; Stock 2006; Demes 2007).

Cross-section geometry of the femur midshaft is principally shaped by the history of mechanical loading, including strains from body mass and weight bearing, muscle action, mechanical stresses, and locomotion (Lieberman, Polk, and Demes 2004; Pearson and Lieberman 2004; Ruff, Holt, and Trinkaus 2006). Additionally nutritional, metabolic, and hormonal status also affect growth (Ericksen 1976; Bridges 1989; Owsley 1991; Bridges 1991a, 1991b, 1995; Stock and Pfeiffer 2001; Mcewan, Mays, and Blake 2005). When bioarchaeologists examine the transverse cross section at the midshaft, they can measure the area of bone tissue (cortical area, total area, or percentage of cortical area), and using engineering beam theory they can estimate second moments of area (proportional to bending strength, Zp) and polar second moments of area (proportional to torsional rigidity, J). Polar second moment of area is calculated as the sum of I_{max} and I_{min}, where I is the distance of the farthest point from the neutral axis of the bone cross section (Currey 2002). These bone geometric properties are a proxy for bone strength and have been used to infer functional adaptations and locomotor and subsistence practices in *adult* skeletons (Ruff et al. 1993; Larsen, Ruff, and Griffin 1996; Bridges, Blitz, and Solano 2000; Pearson 2000; Wescott and Cunningham 2006; Lukacs and Walimbe 2007b).

Increasingly, attention has turned to examining cross-section geometry in subadult long bones. A variety of studies from clinical and biomechanical literature suggest that subadult body mass is the principle strain shaping geometric properties of the midshaft femur cross section. Biomechanics research has repeatedly emphasized the strong relationship between bone mass and body weight during the period of growth and development that occurs prior to adolescence (Van Dermeulen, Beaupré, and Carter 1993; Carter, Vandermeulen, and Beaupré 1996; Moro et al. 1996; Ruff 2003a, 2005a). Several studies have clearly demonstrated that the growth curve for cortical area and the expansion of the periosteal surface parallels the growth curve for body weight in subadults (Moro 1996; Ruff 2005b). Clinical research has also demonstrated that body mass is the primary predictor of midshaft femur strength in subadults and that the addition of other parameters (age, calcium intake, physical activity level) does not improve the predictive results above regressions using body mass alone (Moro et al. 1996). Experimental animal research on the interaction among exercise-induced changes in loading regime, articular surface size, and diaphyseal geometry suggest that the midshaft of the long bones may be highly sensitive to shifts in mechanical loading during growth, even more sensitive than the joint surfaces (articular surfaces) at the bone ends (Lieberman, Devlin, and Pearson 2001).

The mechanical and biological effects of bipedal posture and locomotion on long bones are visible in the midshaft for very young individuals,

beginning at the onset of locomotor behavior after six months postnatal life (Sumner and Andriacchi 1996; Ruff 2003b, 2003a, 2005a; Ruff,
Holt, and Trinkaus 2006). Clearly body mass is the base load to which
the skeleton, particularly the lower limb, is subjected during life (Ruff et
al. 1993; Ruff 2000; Ruff 2002b, 2002a, 2005b). These variables of bone
cross-section geometry, body mass, and locomotion are related to one
another even in young infants and children. These relationships have yet
to be widely applied to studies of locomotor adaptations in archaeological
samples of subadults (Robbins 2007; Cowgill 2008; Robbins and Cowgill 2009; Cowgill 2010; Robbins, Sciulli, and Blatt 2010) but they clearly
demonstrate the ability of immature bone to model in response to fast but
significant changes in the biomechanical milieu.

While there are already several methods available for estimating stature from subadult bones (Telkka, Palkama, and Virtama 1962; Himes,
Yarbrough, and Martorell 1977; Feldesman 1992; Ruff 2007), there are
fewer published methods for accurately estimating body mass from the
subadult skeleton. Ruff (2007) provides methods to estimate body mass
in subadults using the width of the distal metaphysis of the femur in children less than 12 years of age and using the femoral head for older juvenile and adolescent individuals. The formulas based on the bone end
perform well on immature samples of diverse body proportions (Cowgill
and Hager 2007; Cowgill 2008), however, they are not consistently ef-
fective at predicting body mass across the subadult age categories (Ruff
2007). Increasing variance in the scaling relationship between body mass
and the width of the bone ends provides support for the hypothesis that
the width of the distal end of the femur is under strong genetic control
to maintain joint congruence (Ruff et al. 1993; Ruff, Walker, and Trinkaus
1994; Trinkaus, Churchill, and Ruff 1994; Moro et al. 1996; Ruff 1998, 2000,
2002b; Wescott and Cunningham 2006). It follows that high heritabil-
ity will constrain the ability of the bone end to respond to ontogenetic
factors such as body mass. The relationship between body mass and the
midshaft femur geometry also demonstrates increasing variance with age.
The midshaft is plastic and retains the ability to respond to weight bearing
and other activities throughout life. With increasing age, differences in
activity level will play an increasingly large role in shaping cross-section
geometry.

In this chapter I demonstrate a strong scaling relationship between
body mass and a geometric property of the midshaft femur cross section

(femur polar second moment of area (*J*, or torsional rigidity) in subadult skeletons up to eight years of age in a sample from the longitudinal Denver Growth Study (Robbins 2007; Robbins and Cowgill 2009; Robbins, Sciulli, and Blatt 2010). The strong scaling relationship between body mass and femur torsional rigidity is useful to predict body mass in subadult skeletons. In this chapter, I provide a series of age-structured least-squares regression equations to predict body mass from measurements of the midshaft femur in subadult skeletons 0–17 years of age. These formulas were developed and tested using a cadaveric sample from Franklin County, Ohio, research, which was described in more detail in elsewhere (Robbins, Sciulli, and Blatt 2010). In that publication, the formulas to estimate body mass from midshaft geometry were also compared with previously published formulas using the bone ends and results indicated a similar magnitude of accuracy and bias for both of these techniques. The reader is referred to that publication for details of that comparison.

Materials and Methods

A set of age-structured least-squares regression formulas for predicting subadult body mass from femur midshaft cross-sectional geometry (polar second moment of area) were developed using a longitudinal sample of measurements from twenty well-nourished, active juveniles 2 months to 17 years of age selected from a database compiled by the Denver Child Research Council from 1941 to 1967 and used in several previous studies (Ruff 2003a, 2003b, 2005a, 2007). Permission to use these data for this project was obtained from Richard Siervogel, current director of the Lifespan Health Research Center at Wright State University. Ruff measured femur lengths, external diaphyseal diameter, and cortical bone thicknesses (at 45.5% of diaphyseal length) from the Denver sample anteroposterior radiographs (Ruff 2003a, 2003b). Ruff calculated *J* from the diaphyseal external diameter (O'Neill and Ruff 2004). Magnification error was corrected as described previously. An intra-observer measurement error of 3.1 percent for J was reported (Ruff 2007).

The Denver data were originally collected at 2, 4, 6, and 12 months for the first year of life and at six-month intervals through the age of 17 years (although more often annually after age 14 years). Here only data for 2 months (referred to as age category "0") and at annual intervals from 1 to 17 years of age were used to derive estimation equations. Results are intended to apply to individuals at six-month intervals from these ages, for example, the one-year-old formula applies to individuals age 6–17.59 months. The "0" year formula applies to individuals under 6 months of age. Missing data points (2.3% of total sample) were estimated using linear interpolation such that each age category initially contained 20 individuals (following (Ruff 2007). The only exception was the "0" year age category, which contained 15 individuals. Based on comparisons of body mass index, or BMI (weight [kg]/stature [m²]), to national standards (Must, Dalal, and Dietz 1991), one female at age 4–8 and one male at age 6–8 were eliminated as extreme positive outliers, following Ruff (2007). Thus age categories 4 and 5 had a final sample size of 19 individuals and age categories 6–8 had a sample size of 18 individuals.

Least-squares (LS) regression was used to generate age-structured formulas for predicting body mass from polar second moments of area (J). If independent age estimates are available for the target sample, age-structured equations are preferred to account for effects from allometry, as scaling relationships in the size and shape of elements change during growth (Jungers 1988; Ruff and Walker 1993; Ruff, Walker, and Trinkaus 1994; Ruff 2007). Standard errors of the estimates (SEE) were calculated to measure the precision of the predictions for each formula. The percent standard error of the estimate (%SEE) was calculated by dividing the SEE by the mean body mass (kg) for each age category (following Ruff 2007).

Formulas for Body Mass Estimation

Descriptive statistics including mean body mass, standard deviation, and BMI are reported for the Denver sample by age category in Table 5.1. The range of body mass figures recorded for this sample of individuals in age categories 1–17 was 4.5 to 61.5 kg. Age-structured LS regression equations for estimating body mass from the midshaft are provided in Table 5.2. Femur polar second moment of area (J) has a sufficiently strong relationship with and is a statistically significant predictor of body mass in all age categories through infancy and childhood (0–8), and the SEE ranges from 0.27 to 1.75. The method is relatively consistent in its precision with a %SEE = 5.9–7.2 for these age categories. In older age categories (9–17 years), body mass estimates from the midshaft femur are generally less accurate and precise than those from the femoral head, and that measure is thus the preferred method for those age categories (%SEE = 11.9–16.9).

Table 5.1. Descriptive statistics for the Denver Longitudinal Study sample.

Age	n	Weight (kg)	Height (cm)	BMI	J	Distal end (mm)
0	15	4.5	54.9	15	265	-
1	20	9.1	72.8	17	1145	34.3
2	20	11.6	84.9	16	2326	42.5
3	20	13.6	93.9	15	3129	47.1
4	19	15.5	101.6	15	3960	49.4
5	19	17.3	107.6	15	4854	51.3
6	18	19.3	114.7	15	5933	53.7
7	18	21.7	120.4	15	7377	55.7
8	18	24.3	126.1	15	8940	57.5
9	20	28.7	132.9	16	10998	59.9
10	20	31.9	138.6	17	13171	61.9
11	20	35.9	143.2	17	15882	63.8
12	20	39.5	150.2	18	18926	66.0
13	20	44.4	155.3	18	23220	67.9
14	20	49.9	162.3	19	28125	-
15	20	53.9	165.9	20	32587	-
16	20	59.2	170.0	20	38728	-
17	20	61.5	171.2	21	42193	-

Table 5.2. Equations for predicting body mass (kg) from femur torsional rigidity (J) (untransformed data).

Age	Body mass	BMI	Intercept	Slope	F	P	SEE[a]	%SEE[b]
0	4.5	15	3.8	0.003	3.45	0.086	0.27	6.0
1	9.1	17	7.1	0.002	15.40	0.001	0.61	6.7
2	11.6	16	8.1	0.002	16.96	0.001	0.68	5.9
3	13.6	15	10.5	0.001	8.44	0.009	0.92	6.8
4	15.5	15	11.4	0.001	13.45	0.002	1.00	6.5
5	17.3	15	12.8	0.001	14.94	0.001	1.06	6.1
6	19.3	15	14.2	0.001	15.83	0.001	1.23	6.4
7	21.7	15	15.8	0.001	15.10	0.001	1.38	6.4
8	24.3	15	16.0	0.001	19.85	<0.0001	1.75	7.2
9	28.7	16	17.1	0.001	7.43	0.014	4.11	14.3
10	31.9	17	16.3	0.001	8.81	0.009	5.05	15.8
11	35.9	17	18.4	0.001	8.70	0.009	6.06	16.9
12	39.5	18	19.2	0.001	12.24	0.003	6.48	16.4
13	44.4	18	21.1	0.001	16.89	0.001	7.00	15.8
14	49.9	19	30.4	0.001	8.50	0.010	7.29	14.6
15	53.9	20	36.6	0.001	9.46	0.007	6.41	11.9
16	59.2	20	45.8	0.000	3.82	0.067	8.13	13.7
17	59.9	21	46.2	0.000	6.24	0.023	7.84	12.8

Notes: a. SEE = $S_{y \cdot \hat{y}} = \sqrt{\dfrac{\sum (Y - \hat{Y})^2}{n - 2}}$ where Y = observed value of the dependent variable based on the given X, $\hat{Y}$ = predicted value of the dependent variable Y based on the given X, $n-2$ = degrees of freedom for the independent variable.

b. %SEE= SEE / mean body mass (kg) in a given age category.

This result is not unexpected given that we know hormones, activity, and diet play an increasingly large role in bone mass acquisition in older ages. In addition, changes in the shape of the midshaft cross section during adolescence affect the accuracy of estimates for J in these older age categories.

The method presented here also proved to have reliable accuracy when applied to an independent target sample of cadavers from Franklin County, Ohio. The accuracy and bias of the new equations was reasonable (2.5–3 kg and 0.8 kg, respectively), and in fact was similar to the level of accuracy for estimates based on the femoral distal metaphysis in individuals 1–8 years of age (Robbins, Sciulli and Blatt 2010). It is clear that there is a close relationship between bone cross-sectional geometry and body mass given clinical and biomechanical studies which have repeatedly demonstrated a strong relationship between these two variables during growth and development (Ruff and Runestad 1992; Van Dermeulen, Beaupré, and Carter 1993; Carter, Vandermeulen, and Beaupré 1996; Moro et al. 1996; Ruff 1998, 2000, 2002b, 2003a; Pearson and Lieberman 2004; Ruff 2005a; Wescott 2006; Ruff 2007). It is also clear that body mass and activity are not independent in bipedal organisms and both affect the shape of the cross section of the bone and the velocity of bone mass acquisition in the femur beginning early in infancy (Ruff 2003b, 2003a, 2005a).

It is difficult to tease out influences on bone cross sections from body mass, activity levels, muscularity, nutritional status, and hormonal changes, all of which are significant in determining adolescent and adult midshaft robusticity. However, given the synergistic relationship between starvation, hormonal changes, emaciation, and reduced activity levels, it is possible to infer skeletal emaciation rates from extreme reductions in femoral torsional rigidity and use these as a biocultural stress marker. In the next chapter I will apply estimates of body mass (from J) scaled to estimates for stature squared (from long-bone length) to examine population-level differences among the Deccan Chalcolithic villages in the Early and Late Jorwe phases.

6

Reconstructing Health at Nevasa, Daimabad, and Inamgaon

Auxology is an extraordinarily good tool for gauging health both in individuals and in populations. It is a fine-scale gauge, because growth accumulates the effects of the successive small-scale shocks, nutritional deprivations, infections, and emotional and metabolic disturbances to which so many children are subject. It is, of course, non-specific, and it monitors chronic repeated disturbance better than it does a single acute stress, from the effects of which catch-up growth may restore the situation entirely. But it is sufficiently powerful to distinguish, on aggregate, the children of the unemployed or to indicate, in a part of West Africa, the particular season of the year that places a proportion of pregnant mothers at risk of undernutrition.

Tanner (1986:95)

In this chapter, I examine stature and body mass in the skeletal series from Inamgaon to infer biocultural stress levels through time. In the succeeding chapter, I will integrate insights from the demographic profiles and these skeletal growth profiles to examine how subsistence transition and environmental and culture changes impacted infant and child health toward the end of the Deccan Chalcolithic period. Bioarchaeologists are primarily interested in the health and adaptations of past populations. We estimate health and nutritional adequacy using *biocultural stress markers*—measurements of growth disruption and indicators of disease in the human skeleton and dentition (Goodman and Martin 2002). In this context, *stress* is defined as conditions that disrupt normal biological function or homeostasis (Huss-Ashmore 2000). Stressors include extrinsic and intrinsic forces limiting access to the resources necessary to perform basic adaptive tasks (Cohen, Malpass, and Klein 1980; Larsen 1995; Goodman and Martin 2002). Extrinsic stressors in paleopopulations include insufficient food resources, poor sanitation, high parasite load or disease burden, overcrowding, and intra- or intergroup competition for space, privacy, and social resources. Intrinsic stressors include individuals' physiological, metabolic, immunological, hormonal, nutritional, and psychological status.

Manifestations of stress in the skeleton have *biocultural* causes because human biology and culture are inextricably intertwined. Humans are an adaptable species, and environmental challenges are met with behavioral, social, and biological responses that are unique to each population and vary in their effectiveness (Huss-Ashmore 2000). Sociocultural systems, technology, and cultural traditions can serve to buffer populations against stress, but human behavior and social systems are not always adaptive and they do not necessarily develop with a principle goal of buffering all individuals against all stressors all of the time (Goodman and Leatherman 1999). Thus *biocultural stress markers* in the human skeleton represent complex interactions among stresses, susceptibilities unique to each individual, and behavioral and sociocultural factors.

Many biocultural stress markers are *nonspecific*, and different stressors can produce similar manifestations in the skeleton. For example, there are over two hundred identified causes for a commonly used biocultural stress marker known as linear enamel hypoplasia, a disruption of dental enamel formation (Goodman, Armelagos, and Rose 1980). Many stressors are never expressed in the human skeleton because some conditions do not affect bones and teeth and because these tissues are buffered from the effects of stress (Goodman et al. 1988; Ortner 1991b). It is often the chronic and/or severe stressors that are most likely to leave marks on the bones and teeth and skeletons that do have stress markers are also more likely to have co-morbidity issues. Thus it is rare to have certainty about the cause of death for specific individuals in an archaeological context, but the pattern of growth disruption for multiple markers in a population can indicate general *stress levels*. The frequencies of such stress markers can then be compared to estimate the relative success of different strategies employed by human communities to cope with environmental challenges over time.

Stunting (short stature for age) and wasting (low body mass for stature) are *nonspecific* indicators of short-term economic and environmental stress in contemporary populations (Cameron 1979; Bogin and Macvean 1983; Huss-Ashmore and Johnston 1985; Fogel 1986; Tanner 1986; Evelyth and Tanner 1990; Cameron 1991; Steckel 1995; Bogin and Loucky 1997;

Saunders 2000; Floyd 2002; Martin and Goodman 2002; Cameron 2004). Reduced stature and body mass index are associated with nutritional deprivation, parasites, and chronic gastrointestinal issues, particularly diarrhea.

Stature is also commonly used in bioarchaeology to infer biocultural stress levels, and it is clear from these studies that in general, mean adult stature is reduced in populations that rely on agriculture for a large proportion of their subsistence(Genovese 1967; Haviland 1967; Buikstra 1976; Brothwell 1981; Larsen 1982; Cohen 1984; Kennedy 1984; Kennedy, Lovell, and Burrow 1986; Rathbun 1987; Powell 1988; Kennedy et al. 1992; Storey 1992; Larsen 1995; Lambert 2000; Froment 2001; Steckel et al. 2002). This reduction in average adult stature has been attributed to a diet high in a few staple cereals and commensurate reductions in dietary diversity, nutritional insufficiency, and higher probability of food shortage. Declines in adult stature are also commonly attributed to the problem of settling down. Compared to more mobile lifestyles, village life has its own set of problems: socio-sanitation issues, higher prevalence of certain diseases, and increased parasite burden.

Subadult stature is also negatively impacted by the transition to agriculture in many populations globally (Johnston 1962; Armelagos et al. 1972; Lallo 1973; Merchant and Ubelaker 1977; Hummert and Van Gerven 1983; Cohen 1984; Cook 1984; Goodman et al. 1984; Jantz and Owsley 1984; Mensforth and Lovejoy 1985; Owsley and Jantz 1985; Molleson 1989; Lovejoy, Russell, and Harrison 1990; Armelagos, Goodman, and Jacobs 1991; Hoppa 1992; Storey 1992; Saunders and Hoppa 1993; Saunders, Hoppa, and Southern 1993; Cole 1994; Van Gerven, Sheridan, and Adams 1995; Ribot and Roberts 1996; Whittington and Reed 1997; Hoppa and Fitzgerald 1999; Hutchinson 2002; Steckel and Rose 2002).

In this chapter, I use estimates of body mass (from J) scaled to estimated stature squared (from bone length) to examine the prevalence of emaciation in four skeletal samples from the Deccan Chalcolithic. Indices that represent a scaling relationship among bone mass, body mass, and stature have commonly been used to infer *adult* locomotor adaptations, dietary adequacy, and behavior in past populations (Pilbeam and Gould 1974; Trinkaus 1981; Ruff and Walker 1993; McHenry 1994; Ruff 1994; Holliday 1995, 1997; Pearson 2000; Stock and Pfeiffer 2004; Stock 2006). It is clear that the mechanical and biological effects of bipedal posture and locomotion on long bones are also visible in the femur midshaft of very young individuals, beginning at the onset of locomotor behavior after six months postnatal life (Sumner and Andriacchi 1996; Ruff 2003b, 2003a, 2005a; Ruff, Holt, and Trinkaus 2006), and much of the variance in a midshaft cross-section geometric property (J) that approximates torsional rigidity is explained by body mass (see chapter 5). In this chapter, I use estimates of body mass (from J) scaled to estimated stature squared (from bone length) to examine the prevalence of emaciation in four skeletal samples from the Deccan Chalcolithic.

Recently, population level differences have been detected in the scaling relationship between subadult midshaft bone geometry (J), body mass, and stature among archaeological and contemporary samples (Robbins 2007; Cowgill 2008; Robbins and Cowgill 2009; Cowgill 2010; Robbins, Sciulli, and Blatt, in press). For example, when J is scaled to the product of body mass and stature,[1] subadults from a Sudanese archaeological site (Kulubnarti) demonstrated values of J at the low end of the range for human populations, significantly lower than six other archaeological populations (Raymond Dart collection, Indian Knoll, California, Point Hope, Mistihalj, and Luis Lopes) that show no significant differences in growth estimates (Cowgill 2008, 2010). These results indicate poor nutritional status led to reduced body mass for stature and reductions in bone size and strength in the Sudanese population (Robbins and Cowgill 2009).

Significant population level differences in the archaeological populations described above were also detected by an examination of body mass scaled to stature squared (Robbins and Cowgill 2009). Body mass was estimated from J and compared to stature squared estimated from femur length (using formulas in Ruff 2007) to estimate the BMI. The samples from the Dart collection, Indian Knoll, Point Hope, Mistihalj, and Luis Lopes demonstrated a scaling relationship consistent with the contemporary population from the Denver Growth Study. However, the Kulubnarti sample and a sample from Grasshopper Pueblo, Arizona (described in Sumner 1984), demonstrated significant differences in BMI (Robbins and Cowgill 2009). Of 47 infants and children from Kulubnarti, 49 percent had body mass indices below the 95 percent confidence interval for the Denver reference sample (Fig. 6.1). Of 25 infants and children from Grasshopper Pueblo, 55 percent of the population had body mass indices below the 95 percent lower confidence limit of the Denver sample. While BMI ranged from 18 to 21 for the first five years of life in the Denver sample,

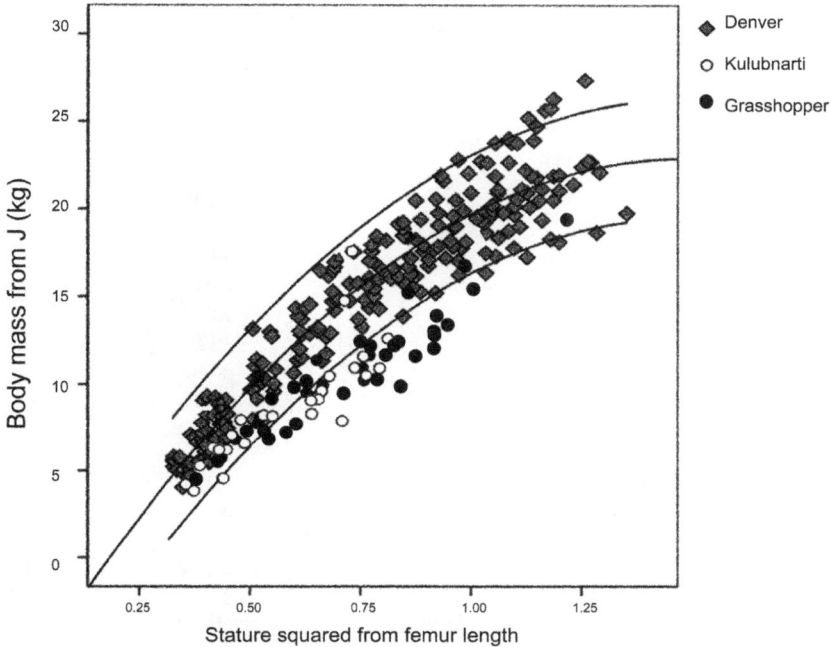

Figure 6.1. Scaling relationship between body mass and stature squared in the Denver sample.

in the prehistoric populations, estimates ranged from 14 to 17 (Table 6.1). Subadults from Grasshopper Pueblo and Kulubnarti had significantly smaller body size and body mass on average relative to the Denver sample, the Dart collection, Indian Knoll, Point Hope, Mistihalj, and Luis Lopes.

Although use of subadult body mass index as a biocultural stress marker has not been fully explored in bioarchaeology, there is a long-standing practice of using midshaft femur cross-section properties (measured as bone thickness or cortical area) as a measure of nutritional status to compare past populations. The compact bone that makes up the walls of the femur midshaft should increase in thickness throughout the subadult lifespan except in circumstances of chronic protein energy malnutrition (Garn 1970). Measures of bone thickness can indicate that bone mass is lower than expected. The problem with using bone thickness or cortical area alone is that these measures do not account for the distribution of bone mass in the cross section, which changes throughout ontogeny

Table 6.1. Body mass index in contemporary and archaeological samples from Denver, Grasshopper Pueblo, and Kulubnarti.

Age (yrs)	Population	n	Mean BMI	Std. dev.
1	Denver	66	18	3.004
	Grasshopper	5	13	2.053
	Kulubnarti	12	14	2.385
2	Denver	37	21	2.205
	Grasshopper	5	14	2.362
	Kulubnarti	7	15	1.858
3	Denver	47	20	2.037
	Grasshopper	5	15	1.011
	Kulubnarti	7	15	1.106
4	Denver	30	19	2.140
	Grasshopper	5	14	0.588
	Kulubnarti	8	15	1.933
5	Denver	44	19	2.099
	Grasshopper	5	17	5.243
	Kulubnarti	7	14	0.984
Total	Denver	224	19	2.659
	Grasshopper	25	15	2.864
	Kulubnarti	41	15	1.776

Source: Robbins and Cowgill (2009).

and has important impacts on bone strength (Ruff, Walker, and Trinkaus 1994).

A similar critique applies to previous studies of percentage of cortical area (%CA = [cortical area/total area] × 100). This measure was also used to estimate nutritional stress in bioarchaeology and it did have one advantage over cortical thickness or cortical area because it is standardized for the total cross-section area, which theoretically allows comparisons between age groups. Thus declines in %CA during the first four years of life were interpreted as evidence for growth disruption in bioarchaeology (Garn 1970; Himes et al. 1975; Cook 1979; Garn 1980; Ashmore 1981; Keith 1981; Hummert 1983; Hummert and Van Gerven 1983; Van Gerven, Hummert, and Burr 1985; Storey 1992; Cole 1994; Mays 1995; Rewekant and Jerszynska 1995; Mays 1999; Rewekant 2001). However, we now know that %CA declines for the first five years of postnatal life (Fig. 6.2) as part of "normal" ontogenetic patterning (Ruff, Walker, and Trinkaus 1994; Carter, Van der Meulen, and Beaupré 1996; Ruff 2003b, 2003a, 2005a). As a measure of bone mass and nutritional stress, %CA is unsatisfactory because it does not distinguish between growth suppression at the

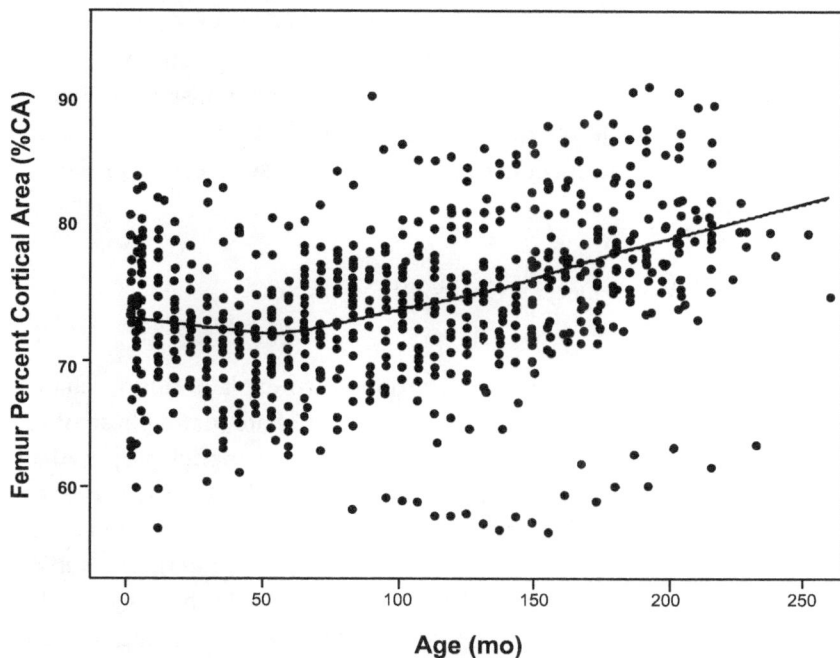

Figure 6.2. %CA at the femur midshaft for age in the Denver sample.

periosteum versus resorption at the endosteum, a surface of active resorption in response to linear growth and modeling pressures (Sumner and Andriacchi 1996; Carter and Beaupré 2001). During postnatal growth, bone tissue is apposited at the periosteal surface as an adjustment to torsional and bending forces (Ruff 1994; Pearson and Lieberman 2004). This offsets losses in %CA due to bone modeling at the endosteal surface because bone mass apposited around the perimeter has a disproportionate effect on bone strength (Biewener and Bertram 1994).

Although %CA is not an efficient indictor of nutritional status or biocultural stress levels, clearly disease and nutritional status have an impact on bone mass, cross-section geometry, body size, and shape (Moro et al. 1996; Van der Meulen et al. 1996; Ruff 2003a; Robbins 2007; Robbins and Cowgill 2009; Robbins, Sciulli, and Blatt 2010). Population differences in bone strength for the lower limb are detectable in subadult skeletons from populations with high levels of biocultural stress such as Kulubnarti and Grasshopper Pueblo. By taking a biomechanical adaptation approach to the problem, reductions in bone mass and bone strength reflect emaciation, including contributions from disease, starvation, and reduced activity levels. Thus comparisons of BMI in age-structured sub-adult skeletal populations can be used as a general, nonspecific marker of biocultural stress levels and can provide information about age of onset, duration, and age-specific hazards for growth suppression in skeletal populations.

Biocultural Stress in Deccan Chalcolithic Samples and Interpretive Models

Two models have previously been proposed to explain culture change and its effect on human populations during the Late Jorwe phase of the Deccan Chalcolithic. The climate-culture change model proposed by Dhavalikar suggests that climate change during the Late Jorwe phase led to abandonment of agriculture and significant negative impacts on human populations after 1000 B.C. (Dhavalikar 1988, 1994). The subsistence transition model proposed by Lukacs and Walimbe suggests that the short-term reversal from agricultural to pastoral/hunting/foraging lifeways during the late Jorwe phase resulted in improvements in infant health status and reduced prevalence of growth disruption (Lukacs and Walimbe 1998; Lukacs and Walimbe 2000; Lukacs and Walimbe 2005a; Lukacs and Walimbe 2007a).

Both models are predicated on a change in diet between the two phases which has largely been confirmed by archaeological and bioarchaeological evidence (Dhavalikar 1988; Dhavalikar et al. 1988; Kajale 1988; Thomas 1988; Lukacs and Walimbe 2007b). The climate-culture change model further suggests that the *cause* of the transition was large-scale climate changes, a prediction that is not supported by recent paleoclimate studies (chapter 2). The subsistence transition model further suggests that the *effect* of the transition was that the Late Jorwe population at Inamgaon experienced a reduction in biocultural stress levels due to greater dietary breadth that accompanied a transition away from relying on barley agriculture toward relying on pastoral, hunting, and gathering subsistence practices. Studies to test hypotheses generated from this model have been inconclusive. The frequency of LEH in the permanent dentition and IPCH in the deciduous dentitions from the Early and Late Jorwe samples at Inamgaon demonstrated no significant differences, when a difference would be expected according to this model.

Examination of the frequency of chronic growth disruption in the enamel formation in primary canine teeth demonstrated a significant difference. The frequency of LHPC is significantly higher in the Early Jorwe Inamgaon sample (47.4%) as opposed to lower frequencies at Nevasa (36.4%), Daimabad (33.3%), and Late Jorwe Inamgaon (35.7%). This result indicates that the infant skeletal sample from Early Jorwe Inamgaon suffered from chronic stresses that were recorded in the dentition. The subsistence transition model predicts that this is a difference between the impact of agricultural versus mixed economies on health. However, because the Early Jorwe sample is an outlier even when compared with other Early Jorwe samples, the result is not entirely consistent with the subsistence transition model as all of these populations were relying on agricultural products as a major component of the diet. One interpretation offered by Lukacs and Walimbe is that smaller sample sizes for Nevasa and Daimabad affected the prevalence of stress markers. An alternative interpretation is that the Early Jorwe sample differs from all other Deccan Chalcolithic samples is one other important way—the demographic profile. Biodemography is one way to evaluate the potential impact of fertility and mortality in shaping the pathological profile.

Osteological Paradox and a Biodemographic Approach to Skeletal Samples

Examination of skeletons to make inferences about human growth is a two-edged sword. While stress markers in individual skeletons can be predictive of growth status at the time of death, comparison among growth profiles in a population of skeletons may provide misleading information because there are lurking effects from demography. Importantly, skeletal series are not a direct representation of health in a once-living population. Instead, they represent individuals who died, and the age structure is a reflection of demographic and epidemiological forces. A sample of skeletons can be affected by an "osteological paradox" (Wood et al. 1992) in which individuals who appear to have numerous stress markers may actually represent "healthy survivors" (Ortner 1991a), whereas individuals that appear healthy from a skeletal standpoint may have succumbed more rapidly to infection, disease, or malnourishment and thus did not have time to express chronic stress in the skeleton. In other words, though it is clear that all of the subadults in these skeletal series died before reaching adulthood, they may represent the sickest, most frail, and most stressed individuals even if they demonstrate no evidence of "stress" because they died before chronic conditions could be expressed in the skeleton (Wood et al. 1992; Hoppa and Fitzgerald 1999). Because paleopathological profiles are shaped by demography, the two studies must be integrated to make inferences about past populations. The only way to deal with the paradoxes of heterogeneous frailty and healthy survivors is to approach the pathological profiles with some knowledge of demography (Saunders and Hoppa 1993) and to distinguish between markers of chronic growth disruption and markers of short-term recent growth disruption.

A biodemographic approach to human skeletal samples addresses both of these goals. The growth of populations is tied to the growth of individuals (Bogin 1999), and mortality, fertility, and subadult growth are all affected by stress levels and the general health of a population. In contemporary and prehistoric populations, infants and children appear disproportionately subject to mortality, morbidity, and growth disruption (Goodman 1989; Cameron 1991). Historically, human populations have had a relatively low life expectancy at birth, and in part this is due to high rates of infant mortality, especially during the first year. The first three years of life are also the most common period during which growth disruption is observed in the skeleton (Saunders and Barrans 1999). Growth velocity is highest during the year following birth, and after the first three years growth becomes canalized such that individuals whose growth has been stunted by poor circumstances tend to remain stunted at least until the rapid velocity of growth during adolescence presents an opportunity to catch-up (Tanner 1986). Because environmental circumstances impact the growth of human populations and the skeletal growth of individuals, I expect to see this relationship in the human skeletal remains. Populations with high-pressure demographic situations, high fertility, and low life expectancy at birth are expected to demonstrate greater frequency of skeletal growth disruption as well.

It is wise to consider whether the results of previous research in bioarchaeology could be consistent with predictions from the osteological paradox (Table 6.2). Daimabad has the lowest rate of LHPC, longest life expectancy, and the lowest fertility rates. Despite lower mortality rates and greater opportunity to express stress, infants have lower rates of dental growth disruption. This result suggests that the population at Daimabad had a lower pressure demographic situation and lower biocultural stress

Table 6.2. Demographic and LHPC profiles for Nevasa, Daimabad, and Inamgaon.

	Total fertility	Life expectancy at birth	% LHPC
Daimabad	3.6	62	33.3
Nevasa	5.2	25	36.4
EJ Inamgaon	8.8	25	47.4
LJ Inamgaon	7.0	12	35.7

Source: Lukacs and Walimbe (2007a).

levels. Whereas the Early Jorwe population at Inamgaon was dealing with
a much more stressful situation with the highest rate of LHPC, low life
expectancy at birth, and very high fertility rates. Clearly the Early Jorwe
population from Inamgaon was stressed.

LHPC rates are lower in the Late Jorwe sample than they are in the
Early Jorwe sample from Inamgaon. This has been interpreted as support for the subsistence transition model, which predicted a reduction
in biocultural stress levels. However, this result is not inconsistent with
Dhavalikar's climate-culture change model if the osteological paradox is
also at work; in other words, lower rates of LHPC can be produced by
lower stress levels or higher mortality rates. The question becomes, are
LHPC rates low in the Late Jorwe phase because the population was less
stressed or because the population was more stressed and infants were
more likely to die before they could express chronic stress? The demographic profile created in chapter 4 allows us to evaluate whether the osteological paradox is a possible explanation for these results. The Late
Jorwe population at Inamgaon had very low life expectancy at birth, so
the results are consistent with expectations of the osteological paradox.

Whereas LHPC is a chronic stress marker primarily reflecting conditions in utero and in the first year of life, emaciation and growth stunting are expected to provide a different kind of information on growth
disruption and consequently the pattern may look quite different. Both
emaciation and growth stunting are used globally as a measure of stress in
infants and children. Emaciation is one of the first effects of starvation, diarrhea, and childhood disease. It reflects acute stress experienced during
the months prior to death. Unlike dental enamel, which is inert once it has
formed, body mass and linear bone growth will respond to changing environmental conditions. Significant "catch-up" growth can occur, particularly in periods like infancy, when the growth trajectory is expected to be

relatively fast (Cameron 2004). Otherwise, children growing up in stressful conditions experience canalization; when the environment remains consistently poor, they may fall far below the expectations for growth and never catch up to their peers. Thus an examination of skeletal growth will demonstrate biocultural stress levels immediately prior to death.

My goal is to understand whether it is high infant mortality or agricultural subsistence that is associated with skeletal growth suppression at Inamgaon. In this chapter, I will first characterize subadult BMI and the prevalence of growth disruption over time in Deccan Chalcolithic samples. Then I will examine whether emaciation rates vary by subsistence category or by demographic parameters. To test the hypothesis that agricultural populations differed from populations relying on a more mixed economic system, I will compare pooled Early Jorwe samples from Inamgaon, Nevasa, and Daimabad with the Late Jorwe sample from Inamgaon. A significant difference between these samples would provide support for the subsistence transition model proposed by Lukacs and Walimbe. Such a result would indicate that despite the differences in life expectancy at birth and fertility rates among the Early Jorwe samples, subsistence was the driving force shaping pathology profiles. Such a result would indicate that Early Jorwe samples in general were stressed due to reliance on cereal foods and the Late Jorwe sample benefitted from the subsistence transition in regard to rates of emaciation.

To test the hypothesis that rates of growth suppression are influenced by demographic parameters, I will examine growth suppression in BMI over time at Inamgaon in light of information about infant mortality rates. The Late Jorwe at Inamgaon was characterized by relatively low life expectancy at birth. If this sample also demonstrates increased rates of emaciation, this result would support Dhavalikar's hypothesis that the Late Jorwe population at Inamgaon was more stressed than the Early Jorwe population. If the Late Jorwe sample demonstrates lower frequency of emaciation, this result would support the subsistence transition model— that reduced reliance on agriculture led to improvements in health status despite the demographic collapse. However, the latter result could also indicate that the osteological paradox has affected the pathological profile for skeletal and dental growth disruption.

Materials and Methods

I developed expectations for the scaling relationship between body mass and stature squared (that is, body mass index) using data from a contemporary reference population, the database compiled by the Denver Child Research Council from 1941 to 1967. This longitudinal sample was comprised of 20 well-nourished, active infants and children, 10 males and 10 females of "white" European ancestry. Measurements were obtained at 2, 4, 6, and 12 months for the first year of life and at six-month intervals through the age of 17 years. Two hundred and twenty-three measurement events, those occurring between 2 to 60 months of age, were included in this study. I constructed age categories 1–5 in yearly intervals centered on the whole year; for example, age category 1 includes individuals 0.5 to 1.49 years. Age category 0 includes individuals measured at 2 months postnatal life.

Data included in this analysis were age, weight, stature, femur diaphyseal length, midshaft cortical bone thickness at the medial, and lateral wall of the midshaft. Chris Ruff calculated torsional rigidity (J) for this sample using the ellipse model method (EMM) from the biplanar radiograph data (O'Neill and Ruff 2004), corrected for magnification (Maresh 1959, 1970; Ruff 2003a, 2005a, 2007). The polar moment of area (J) was calculated as "/32 $\times$ (T^4-M^4), where T = total mediolateral diameter and M = (total breadth–summed medial and lateral cortical breadths) (Ruff 2007). This method bases the reconstruction of bone geometry on data from biplanar radiographs by assuming the bone cross section is roughly circular. Estimates of polar second moments of area (J) using EMM may be overestimated in older individuals, and LCM, which accounts for the bone's contours, is preferred in samples of older juveniles and adults. However, for young infants and children the two methods produce small differences in estimates of geometric properties (O'Neill and Ruff 2004).

An evaluation of body mass index requires complete long-bone lengths to estimate stature, intact femur midshafts to estimate body mass, and an independent estimate for age at death (from dental development and eruption timing). Using methods for age estimation described in chapter 4, I determined that there were a total of 269 subadults (89.4% of the total number of 301 skeletons) from the Early and Late Jorwe phases at Nevasa, Daimabad, and Inamgaon (see Appendix B for details). I limited the sample for this analysis to individuals less than five years of age because

Table 6.3. Sample size and proportion of individuals with dental age estimates, complete femur lengths, and intact femur midshaft compact bone.

| | Inamgaon | | | | | | | |
| | EJ | | LJ | | Nevasa | | Daimabad | |
Age (yrs)	n	Prop.	n	Prop.	n	Prop.	n	Prop.
≤ 1	8/25	0.32	15/58	0.26	3/30	0.10	6/19	0.32
2	0/4	0.00	0/11	0.00	2/22	0.09	0/8	0.00
3	0/9	0.00	2/7	0.29	0/5	0.00	0/0	0.00
4	0/0	0.00	0/12	0.00	0/7	0.00	0/4	0.00
5	0/1	0.00	2/4	0.50	0/1	0.00	0/1	0.00
Sample / total N ind. 0–5 yrs	8/39	0.21	19/92	0.21	5/65	0.08	6/32	0.18

there were few individuals in the older subadult age categories and age structured population comparisons would not be sufficiently powerful to detect significant population differences. There were 232 individuals age 0–5.49 years (86.2% of all subadults). Of this sample of young infants and children, 112 individuals had complete long-bone lengths available to estimate stature, but only 62 individuals had complete long-bone lengths and dental age estimates available (Appendix C). There were 59 subadult individuals with intact femoral midshafts (Appendix D); only 38 individuals from the Deccan Chalcolithic samples (16.4% of the subadult sample ≤ 5 years of age) had intact femur midshafts, complete long-bone lengths, and dental age estimates available (Appendix E).

I divided Chalcolithic samples into age categories such that the perinate category (category 0) includes individuals ≤ 3 months old at death. Age categories 1–5 were constructed in yearly intervals centered on the whole year. For example, age category 1 included individuals 0.5 to 1.49 years. Table 6.3 provides details about the age structure of the sample available for analysis in age categories up to 5.49 years old. These samples are primarily comprised of infants who died within the first 1.49 years of postnatal life. The samples size were small: 5 individuals from Nevasa (7.0%), 6 individuals from Daimabad (18.2%), 8 individuals from the Early Jorwe phase at Inamgaon (21%), and 19 individuals from the Late Jorwe phase at Inamgaon (29%).

I measured long-bone diaphyseal length for unfused femurs and estimated stature using formulas from Fazekas and Kósa (1978) for perinatal individuals less than 44 lunar weeks of age (1 postnatal month). I estimated stature in older infants and children using regression formulas developed from the Denver Growth Study data (Maresh 1970). If more than one bone was available, I used the length of the bone that provided the most precise prediction of stature according the regression formula standard error (in descending order: femur, humerus, tibia, radius, ulna, or fibula).

I radiographed long bones at the offices of Dr. Ram Tapasvi in Pune. Data collected for this analysis included midshaft mediolateral diameter and cortical thickness at the medial and lateral wall of the midshaft femur. Midshaft dimensions were measured at 50 percent diaphyseal length. I calcluated polar second moments of area from the radiographs using the EMM, a method described in chapter 5. I calculated torsional rigidity, or J as "/32 × $(T^4–M^4)$, where T = total mediolateral diameter and M = (total breadth–summed medial and lateral cortical breadths). I calculated body mass using age structured equations provided in chapter 5. In this analysis, I used stature squared as the scaling factor to compare estimates for body mass (BMI = body mass [kg]/stature [m^2]).

The age structure of the samples used in this analysis is shown in Table 6.3. The first comparison examines whether there are significant among the samples from Nevasa (N = 5), Daimabad (N = 6), Early Jorwe Inamgaon (N = 8), and Late Jorwe (N = 19) Inamgaon. The second comparison is between a pooled sample from the Early Jorwe phase (Early Jorwe Inamgaon, Nevasa, and Daimabad; N = 19) and the Late Jorwe sample from Inamgaon (N = 19) to test for a correspondence between subsistence strategy and growth disruption. The last comparison is of the sample from the Early Jorwe phase at Inamgaon (N = 8) and the Late Jorwe at Inamgaon (N = 19). Because sample sizes are particularly small in age-structured analyses of archaeological samples, I tested the significance of population differences and the timing of the emergence of these differences using nonparametric Kruskal-Wallis tests. I used a nonparametric Mann Whitney U test to examine which samples differed.

Population Level Differences in Stature and Body Mass

Descriptive statistics for stature (Table 6.4), body mass (Table 6.5), and body mass index (Table 6.6), and box plots for stature (Fig. 6.3), body mass (Fig. 6.4), and body mass index (Fig. 6.5) show the characteristics of body size and shape for each sample. The first research question was: Do

Table 6.4. Mean estimated stature.

	Age category											
	0		1		2		3		4		5	
Population	Mean	Std. dev.	Mean	Std. dev.	Mean	Std. dev.	Mean	Std. dev.	Mean	Std. dev.	Mean	Std. dev.
Denver	57.91	3.37	68.75	4.57	81.59	4.00	91.79	3.69	99.64	4.37	106.43	4.34
EJ Inamgaon	61.43	9.91	74.68	7.99	78.70	.	93.40	6.93	92.66	6.26	104.50	.
LJ Inamgaon	61.98	12.21	81.48	4.98			84.90	2.70	87.25	3.32	96.00	2.33
Nevasa	59.70	13.15	75.76	7.01	76.50	7.07	85.30	6.22				
Daimabad	64.00	.	79.17	2.40								

Table 6.5. Mean estimated body mass.

	Age category											
	0		1		2		3		4		5	
Population	Mean	Std. dev.	Mean	Std. dev.	Mean	Std. dev.	Mean	Std. dev.	Mean	Std. dev.	Mean	Std. dev.
Denver	5.33	0.88	8.14	1.16	10.89	1.12	13.14	1.13	14.99	1.49	17.06	1.63
EJ Inamgaon	4.11	0.43	8.15	0.80								
LJ Inamgaon	4.61	1.53	8.47	0.57			12.84	0.83			15.85	0.39
Nevasa	.	.	9.06	0.68	11.28	2.00						
Daimabad	4.81	1.14										

Table 6.6. Mean estimated body mass index.

	Age category											
	0		1		2		3		4		5	
Population	Mean	Std. dev.	Mean	Std. dev.	Mean	Std. dev.	Mean	Std. dev.	Mean	Std. dev.	Mean	Std. dev.
Denver	16	0.203	17	0.197	16	0.166	16	0.146	15	0.153	15	0.139
EJ Inamgaon	16	0.693	17	1.329								
LJ Inamgaon	15	1.078	14	0.785			11	5.29	15			0.162
Nevasa	.	.	14	0.874	19	0.096						
Daimabad	14	2.063										

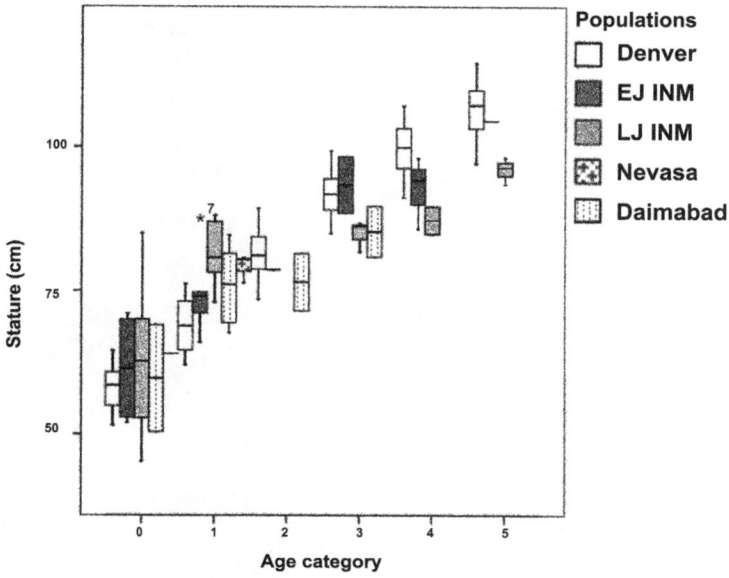

Figure 6.3. Box plots for stature in the Denver and Deccan Chalcolithic samples.

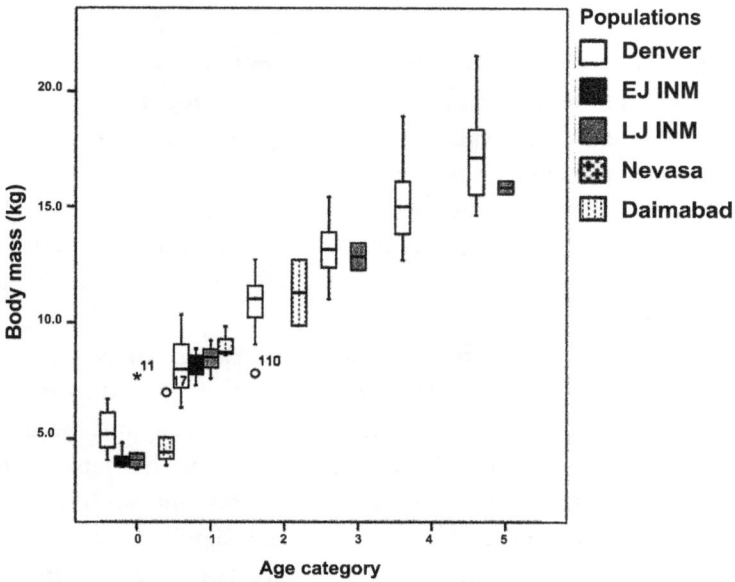

Figure 6.4. Box plots for body mass in the Denver and Deccan Chalcolithic samples.

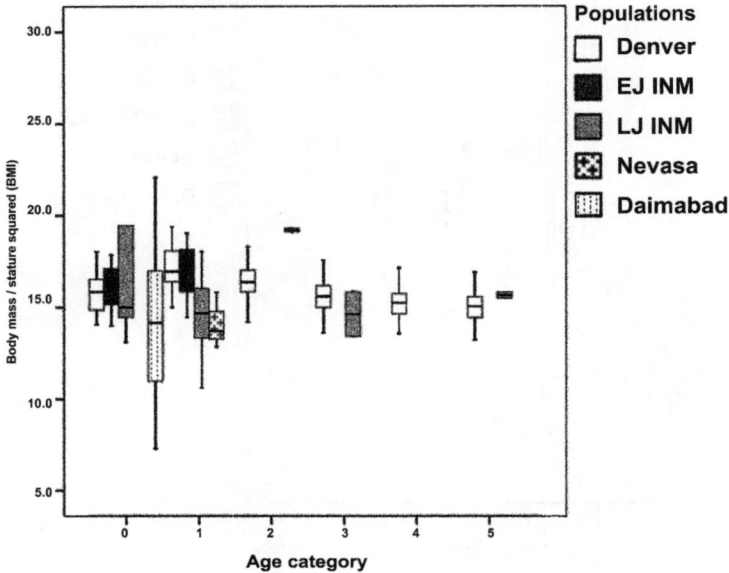

Figure 6.5. Box plots for body mass scaled to stature squared in the Denver and Deccan Chalcolithic samples.

the Deccan Chalcolithic samples differ significantly from one another? Because of the significant differences in both the demographic and the dental growth profiles, my hypothesis was that these samples will demonstrate significant differences from one another and from the reference sample. This hypothesis was supported. In a nonparametric Kruskal-Wallis test, there were significant differences between samples for stature (χ^2 = 25.508, P ≤ 0.001), body mass (χ^2 = 33.675, P ≤ 0.001), and body mass index (χ^2 = 13.193, P = 0.010).

I used pairwise Mann-Whitney U tests to determine which samples differed significantly from expectations using pooled age categories 0 through 5 (Table 6.7). Results demonstrate that Daimabad and the Late Jorwe sample from Inamgaon differed significantly from the Denver reference sample in stature, body mass, and body mass index. The Early Jorwe samples from Inamgaon and Nevasa were not significantly different from expectations based on the Denver reference sample. An examination of the residuals for body mass index demonstrates that 33.3 percent of individuals from Daimabad and 15.8 percent of individuals from the

Table 6.7. Mann Whitney test for significant differences among the Deccan Chalcolithic samples (pooled sample of age categories 0–5).

		Inamgaon				
		Denver	EJ	LJ	Nevasa	Daimabad
EJ Inamgaon	Stature	≤ 0.001*				
	Body mass	≤ 0.001*				
	BMI	0.564				
LJ Inamgaon	Stature	0.016*	0.025*			
	Body mass	0.003*	0.097			
	BMI	0.004*	0.130			
Nevasa	Stature	0.298	0.005*	0.638		
	Body mass	0.230	0.010*	0.150		
	BMI	0.951	1.0	0.446		
Daimabad	Stature	0.004*	0.755	0.199	0.095	
	Body mass	≤ 0.001*	1.0	0.080	0.008*	
	BMI	0.021*	0.073	0.218	0.310	

Note: *. Significant at $P \leq 0.05$.

Late Jorwe Inamgaon sample had strongly negative values that fell below the 95 percent confidence interval for BMI (Fig. 6.6). All of the individuals from the Early Jorwe phase at Inamgaon fell within the 95 confidence interval for BMI.

The samples are heavily skewed toward representation of the youngest infants (≤ 1.49 years of age), and the most significant differences between samples are located in age categories 0–2 (individuals aged 0–2.49 years). When body mass index is plotted relative to age, it is clear that growth faltering in the archaeological samples occurs in age categories 0–2 (Fig. 6.7). Body mass index was significantly different in age category 0 (χ^2 = 8.40, P = 0.038), category 1 (χ^2 = 14.363, P = 0.002), and category 2 (χ^2 = 5.514, P = 0.019) using a Kruskal-Wallis test. Mean body mass index for infants ≤ 2.5 years old at Daimabad and Late Jorwe Inamgaon was 14.5. Body mass index in those age categories at Nevasa and Early Jorwe Inamgaon matched the Denver reference sample (mean = 16.5). Older age categories had small sample sizes and differences were not statistically significant. Young infants ≤ 12 months of age from Early Jorwe Daimabad and Late Jorwe Inamgaon differed from the expectations for BMI developed from a contemporary reference sample; growth profiles for Early Jorwe Inamgaon and Nevasa were not significantly different from the contemporary reference sample.

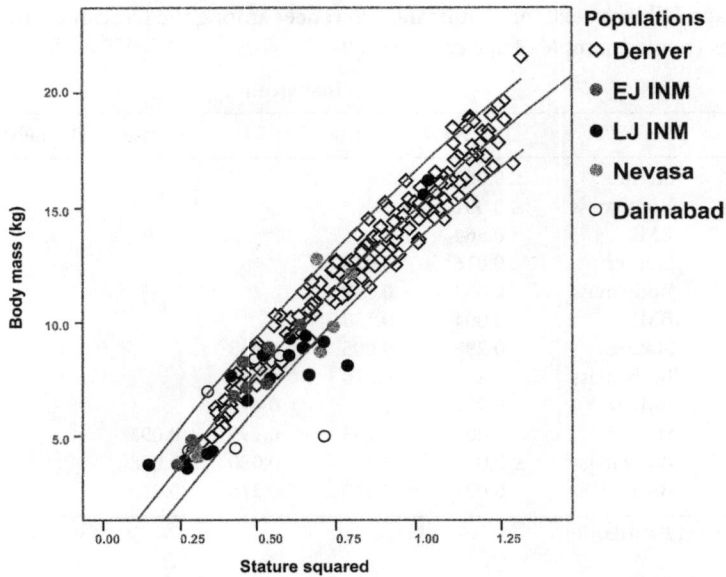

Figure 6.6. Scatter plot of body mass scaled to stature squared.

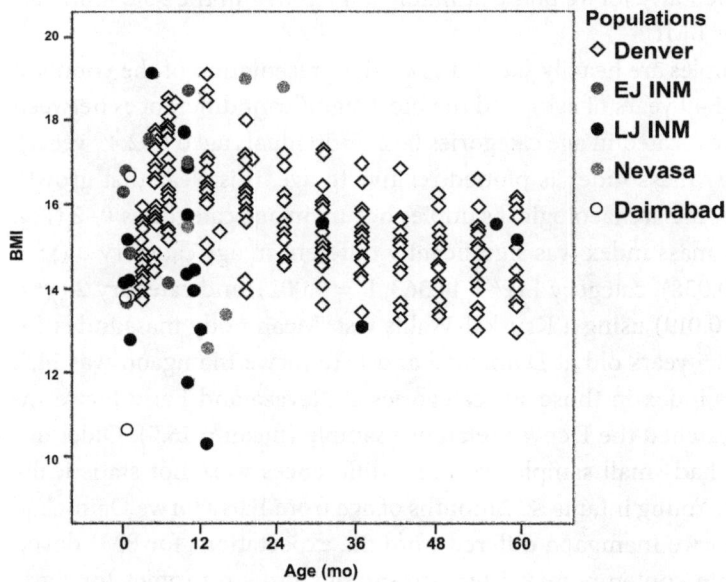

Figure 6.7. Body mass index plotted against age.

Table 6.8. Mann Whitney tests for significant differences among samples based on differences in subsistence (infants age 0–1.49).

| | Subsistence transition model (LJ Inamgaon vs. pooled EJ sample) | | |
	Stature	Body mass	BMI
U	71.00	77.00	94.00
Z	-1.221	-0.944	-0.161
P	0.235	0.363	0.892

Notes: Significant at $P \leq 0.05$.

The second research question was: Do Early Jorwe samples differ significantly from the Late Jorwe sample? The subsistence transition model suggests that biocultural stress levels vary by subsistence activity and thus predicts that there will be a significant difference between Early Jorwe agricultural populations and Late Jorwe Inamgaon. The Late Jorwe sample will differ significantly from the agricultural samples of the Early Jorwe phase because biocultural stress levels are primarily reflecting differences in subsistence. I used a Mann Whitney U test to examine differences between the pooled Early Jorwe samples and the Late Jorwe sample from Inamgaon (Table 6.8). For this comparison, I limited the samples to age categories 0 and 1 because these age groups were well represented in all four samples and the previous analysis demonstrated that any significant difference among these samples was in these age categories. There was no significant difference between the pooled Early Jorwe samples versus the Late Jorwe sample. The hypothesis that population differences are related to broad subsistence categories is not supported.

The third research question was: Are rates of growth suppression higher during the Early or the Late Jorwe at Inamgaon? The subsistence transition model predicts that the Late Jorwe population was less stressed and will demonstrate lower prevalence of growth disruption than the Early Jorwe because they reduced their reliance on agriculture. An examination of mean stature and mean body mass reveals that these values are low for infants (age categories 0 and 1) in the sample from the Early Jorwe at Inamgaon (Tables 6.4 and 6.5). However, when mean body mass index is examined, the Early Jorwe infant sample from Inamgaon had significantly greater BMI than any other sample—Early Jorwe Nevasa, Daimabad, and Late Jorwe Inamgaon (Table 6.6). The Early Jorwe infants were not different from the contemporary reference sample (Fig. 6.6). Emaciation rates are highest in the Late Jorwe sample from Inamgaon. The hypothesis that abandoning agriculture led to reductions in biocultural stress levels is not supported. Rather, when the people of Inamgaon could no longer rely on agriculture, stress levels increased. The Late Jorwe at Inamgaon had high fertility, very high infant mortality, and greater prevalence of skeletal growth suppression. My results from all of these comparisons are conservative. Although sample sizes are small, they are statistically significant using nonparametric tests, which have limited power to detect differences, even when they really exist.

Conclusions

Both dental and osseous markers demonstrate significant differences existed among the Deccan Chalcolithic populations. The Early Jorwe sample from Inamgaon has the highest frequency of LHPC and the lowest frequency of skeletal emaciation. These two stress markers might differ in frequency because BMI is a marker of growth disruption in the months before death, whereas LHPC is a marker of chronic growth disruption. Alternatively, they may differ because of heterogeneous frailty or because the prevalence of LHPC was affected by an osteological paradox. High fertility and low life expectancy at birth during the Late Jorwe phase may have led to lower rates of enamel defects because infants died before they could express this chronic stress marker. However, there are other possible explanations. The LHPC rates may have been higher during the Early Jorwe at Inamgaon because reliance on agricultural foods created micro-nutrient deficiencies that specifically predispose individuals to plane-form enamel defects (e.g., hypovitaminosis A) or the high prevalence could be related to tooth and jaw size differences during the Early Jorwe phase (Skinner and Newell 2000; Lukacs, Walimbe, and Floyd 2001; Lukacs 2009). The Deccan Chalcolithic sample sizes are small and conclusions must be regarded with some caution. In both studies of teeth and of long bones, however, it is the Early Jorwe at Inamgaon that stands apart as an outlier, and the pathological profiles do appear to have been shaped by both differences in both demography and diet.

The subsistence transition model had two components: dietary changes occurred between the Early and Late Jorwe at Inamgaon and the Late Jorwe benefitted from reduced reliance on agriculture. Several studies have found evidence for dietary change through time. However, the hypothesis that the Late Jorwe population experienced an improvement in biocultural stress levels is not supported by my analysis. The difference between samples segregated by subsistence type was not statistically significant.

The climate-culture change model also had two main components. Dhavalikar predicted dramatic and dire consequences for health in the Late Jorwe, given archaeological evidence that poverty and starvation would have affected the people who remained at Inamgaon. My results support this part of the model. The Late Jorwe at Inamgaon did not represent an improvement in circumstances. However, in chapter 2 I demonstrated that the changes in culture and the subsistence transition were not driven by large-scale climate changes. The paleoclimate evidence indicates that semiarid conditions were well established in South Asia before the Deccan Chalcolithic began. In chapter 7, I develop a new biodemographic model for understanding the Early and Late Jorwe phases of the Deccan Chalcolithic.

This study is the first systematic comparison of subadult BMI to document differences in emaciation from both archaeological and contemporary samples. Previous studies have demonstrated population differences in values for torsional rigidity scaled to the length of the femur (Robbins 2007; Robbins 2008; Cowgill 2009; Robbins and Cowgill 2009) and established that variation is present in the biomechanical signatures of archaeological populations beginning early in infancy. The analysis I present in this chapter and chapter 5 suggests that the variation in subadult biomechanical signatures is largely explained by subadult body mass. Despite an expectation that there will be strong and predictable scaling relationships among torsional rigidity, body mass, femur length, and stature in contemporary and archaeological populations of subadults (Ruff 2003a, 2003b, 2005; Cowgill 2008; Cowgill 2009; Robbins and Cowgill 2009), a significant proportion of individuals in the Deccan Chalcolithic samples (15.8%) have BMI values significantly below the 95 percent confidence interval. This result, and previous results from Grasshopper Pueblo (52%) and Kulubnarti (66%), confirm that emaciation can be detected in the human skeleton and can be compared to infer stress levels in past populations.

While my results help to distinguish between models for culture change in the Deccan Chalcolithic period of Indian prehistory, several issues require further studies on subadult samples to make more specific predictions about the impact of stress on the subadult skeleton and its predictive power to infer biocultural stress levels in prehistoric populations. We should develop ontogenetic perspectives on biomechanical adaptation. It is necessary to examine the issue of emaciation in additional archaeological populations to examine population differences in compact-bone properties throughout the lifespan and their relationship to locomotor, settlement, and subsistence behaviors. Although our work on the samples from the Dart collection; Indian Knoll, California; Point Hope, Alaska; Mistihalj; and Luis Lopes demonstrates that the majority of populations conform to expectations about body mass for stature (Robbins and Cowgill 2009; Cowgill and Robbins, in prep.), we have yet to comprehensively explain the differences among the samples from Grasshopper Pueblo, Kulubnarti, and the Deccan Chalcolithic which differ significantly from expectations (Cowgill and Robbins, in prep.).

Another important line of research will be to integrate bone geometry with information about bone biology during ontogeny. Recently, scholars across disciplines have recognized the potential importance of looking at bone cross-section properties, bone mass, and histology across the lifespan (Bianchi 2005, 2007; Cooper et al. 2002; Duan et al. 2003; Eisman et al. 1993; Gluckman and Hanson 2004, 2005; Hui et al. 1990; Kornak and Mundlos 2003; Lane 2006; Loro et al. 2000; Oreffo et al. 2003; Ott 1990; Oliver et al. 2007; Parfitt 1994; Parfitt et al. 2000; Schoenau 1997). However, these studies have yet to be fully integrated into an examination of the relationship between cross-section and histological properties for individuals within and outside the "normal" range of variation.

The bone that we acquire as children affects the size, shape, and microstructure of our bones as adults. Scaling relationships are not static as bone is a dynamic tissue—it grows and changes shape through modeling, and the microstructure is reorganized by remodeling throughout the lifespan. Through these contingent processes, bone geometry and microstructure actually represent a life history of functional adaptations and nutritional, hormonal, and metabolic constraints during growth and development. Research on the relationship between subadult cross-section properties and bone histology could provide a more detailed reconstruction of growth, growth disruption, and strength parameters using information about the quantity and distribution of tissue types, BSUs,

Haversian surfaces, and mineralization differences. These future directions will potentially provide bioarchaeologists with additional tools for understanding bone biomechanics and the derangement of the growth process in archaeological populations in different environmental, locomotor, subsistence, and sociocultural circumstances.

7

◇◇◇◇◇◇◇◇◇◇◇◇◇◇◇◇

Conclusion

Animal populations and human populations as well are characteristically extremely conservative in maintaining adaptive postures and structures which have proved successful. A long and well-studied paleontological record suggests that in the biological world, change does not occur for its own sake but rather results from altered selective pressures which necessitate adjustive modifications on the part of the population involved. If . . . [a subsistence practice] was such a successful mode of adaptation over such a long period of time, and if most human populations are as conservative as anthropologists have observed them to be, we are faced with answering the question why this form of adaptation was ever abandoned.

Cohen (1977:1)

What does the combined evidence from archaeological, paleoecological, paleoclimate, and biodemographic studies tell us about the human populations and environmental conditions in the Early and Late Jorwe at Chalcolithic Inamgaon? What significant differences occurred about 1000 B.C. that led to abandonment of the majority of most villages in west-central India? How did Inamgaon persist into the Late Jorwe phase when other settlements were abandoned and why was it too eventually abandoned? In chapter 1, I summarized the interpretive models developed to explain culture change in the Deccan Chalcolithic: the climate-culture change model presented by Dhavalikar (1988) (as modified by Panja [1996, 1999, 2003]) and the subsistence transition model presented by Lukacs and Walimbe (1998, 2000, 2005, 2007a, 2007b). In this chapter, I propose an alternative model for understanding the Deccan Chalcolithic, the biodemography model.

The climate-culture change model, the subsistence transition model, and the biodemography model all agree that there was population expansion and reliance on agriculture in the Early Jorwe phase (Fig. 7.1). Archaeological evidence specifically demonstrates population expansion

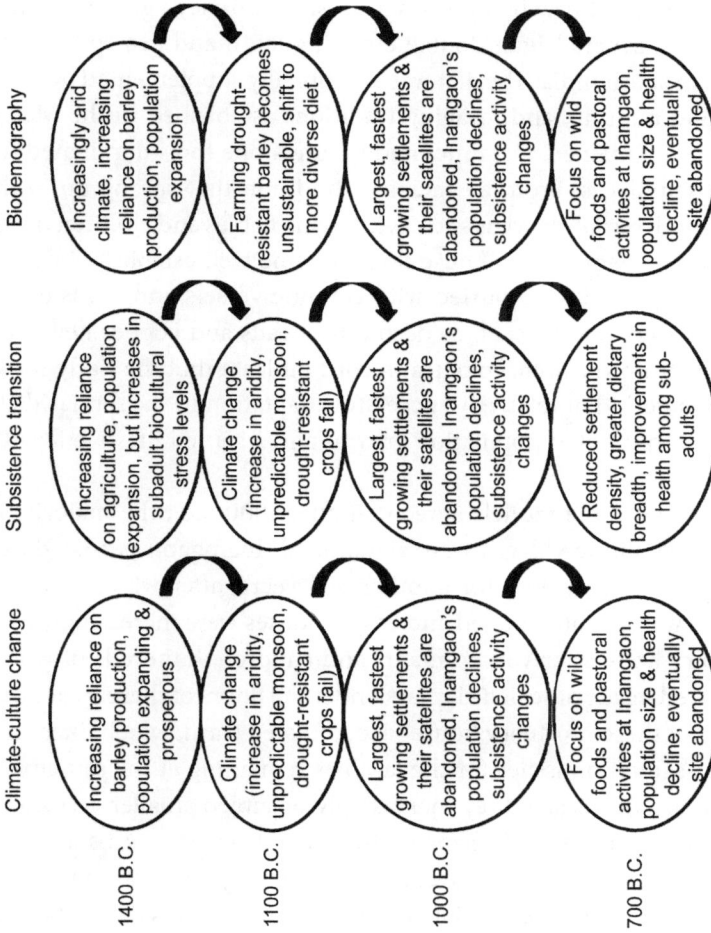

Figure 7.1. Summary of models to explain culture change in the Jorwe phase of the Deccan Chalcolithic.

and settlement growth for Daimabad and Inamgaon. Daimabad grew from 20 to 30 ha in area during this time. Inamgaon grew from a very small settlement in the incipient Malwa phase to an area of 5 ha in the Early Jorwe phase. Families built large rectangular dwellings to accommodate the growing number of people, and to feed the growing population, the communities relied heavily on drought-resistant barley agriculture and, increasingly, farmers began double cropping lentil and pea species imported from south India. The diet was additionally supplemented by cattle keeping and hunting, and foraging traditions established in the Malwa phase were maintained. Ceramics and groundstone tools dominated the artifact assemblages. From an incipient tradition that sprang up in the Malwa period, burial customs such as interring babies and children under age five in jars beneath the house floor became well established during this time. The dead were buried with ceramic vessels and occasionally other utilitarian objects, such as decorative beads and copper items. All indications from the archaeological record indicate that the Early Jorwe phase was a period of relatively successful adaptation to the semiarid climate, which promoted population growth and a settled lifestyle for 400 years.

Similarly, all three models agree that after 1000 B.C., the majority of Deccan Chalcolithic villages were abandoned. Inamgaon persisted as a smaller, poorer settlement for another 300 years, after which much of this village was also abandoned and empty houses were increasingly used as mortuary houses for young infants and children, hundreds of whom were interred inside of sealed clay jars under the floors of these structures. The floors of these mortuary houses accumulated trash, which likely created socio-sanitation issues for the families remaining at the settlement. As family size decreased, they increasingly occupied smaller, less costly round pit dwellings. Other changes also suggest poverty and social disruption during this period, including a decline in agricultural subsistence activities that reduced the absolute volume of food produced and changed the balance of subsistence efforts devoted to other activities. Drought-resistant barley was no longer the focus of subsistence efforts in the dry season and instead the diet was increasingly supplemented by foraging for Indian jujube and freshwater mussels. Monsoon-season crops were still grown, including saline-tolerant lentils (Kajale 1988).

Finally, all three models also agree that Late Jorwe people shifted their efforts from cattle keeping to sheep/goat herding. Cattle were still kept but were raised to older ages, suggesting a shift in the function or the meaning of the cow (Pawankar 1996). Sheep and goats were slaughtered at young ages and used for food. The number of antelope species represented and the proportion of their remains in the food refuse did not change between the Early and Late Jorwe phases, indicating that people were unable to exploit this resource more heavily when other food sources failed. Freshwater mussels and Indian jujube, both considered starvation foods today, were exploited very heavily in the Late Jorwe phase, and the proportion of these items in the food refuse increases over time. This indicates an increased reliance on these foods toward the end of the site's occupation.

While there is evidence for continuity in burial traditions throughout the Jorwe phases (Raczek 2003), other features of the material culture changed significantly after 1000 B.C. In the Late Jorwe phase, ceramics were built from coarse clay bodies and were largely undecorated. Groundstone tools for processing grain and vegetal foods were less common, and the population increasingly focused on producing chipped-stone tools, projectile points, and other forms associated with hunting activities. As freshwater mussels began to predominate in the food refuse, shells were also increasingly used to create decorative objects. The presence of coastal species indicates that Late Jorwe people may have participated in new trade networks across increasingly far distances or may have increased their participation in established trade networks. Copper objects were never common, but they were less common in the Late Jorwe phase and increasingly represented utilitarian items.

There are also major points of disagreement among the three models. I disagree about the cause of the changes at the end of the Early Jorwe phase of the Deccan Chalcolithic. Both the climate-culture change and the subsistence transition models are based on the premise that large-scale climate changes were responsible for culture change and subsistence transition at Inamgaon. This conclusion came out of the original interpretations of lake core evidence interpreted by Singh and colleagues (Singh, Joshi, and Singh 1972; Singh, Chopra, and Singh 1973; Singh et al. 1974; Singh, Wasson, and Agrawal 1990) and studies of fluvial dynamics (Kale and Rajaguru 1987, 1988; Kale 2007), which strongly suggested that climate change was associated with and could be responsible for culture change in South Asia. Singh and colleagues inferred the following climate sequence from Rajasthan lake cores: a period of wetter climate from 3000 to 1700 B.C., more arid conditions from 1700 to 1500 B.C., a relatively wet phase from 1500 to 1000 B.C., then arid conditions returning until 500 B.C. From this sequence it would appear that wetter phases of climate were associated with settled agrarian life and urbanization in the mature phase of the Indus civilization, Early Jorwe phase of the Deccan Chalcolithic, and the Early Historic period. Arid phases coincided with periods of social disintegration in the archaeological record.

A thorough review of the most recent paleoclimate evidence (chapter 2) demonstrates that the semiarid climate regime was established in South Asia by 3000 B.C. and persisted without major interruptions throughout the Deccan Chalcolithic period (Bryson and Swain 1981; Swain, Kutzbach, and Hastenrath 1983; Wasson, Smith, and Agarwal 1984; Enzel et al. 1990; Roberts and Wright 1993; Von Rad 1999; Jain and Tandon 2003; Prasad and Enzel 2006; Sinha et al. 2006; Roy et al. 2008). This has broad implications for South Asian prehistory in general because it indicates that sedentism and even urbanization are not necessarily associated with wetter climate phases in South Asia. The evidence overwhelmingly suggests that civilization in South Asia has flourished during semiarid periods and that population growth reached its most rapid pace in prehistory, when large populations were fed by dry plough agriculture focused on drought-resistant cereal production.

The paleoclimate records broadly indicate reductions in total rainfall in the second millennium B.C., but they do not indicate whether rainfall occured on a more or less predictable schedule. Climatic uncertainty and unpredictability can have a powerful effect on human communities. It is highly likely that there were small fluctuations in the predictability of the rainfall or even short-term catastrophic events or periods of cooling temperature that may have had negative impacts on society in South Asia during this time but which cannot be resolved in the paleoecological record collected thus far (Baillie 1991; Pilcher and Hall 1992). Although there may have been short-term fluctuations in the weather, there is no indication of a significant shift that would cause the dramatic culture change we see in the Late Jorwe phase. In fact, some studies indicate the aridity peaked at 1500 B.C., just prior to the beginning of the Early Jorwe phase of the Deccan Chalcolithic. It is clear, then, that large-scale climate change was not the driving force behind the abandonment of Early Jorwe settlements and culture change in the Late Jorwe phase of the Deccan Chalcolithic.

Large-scale climate change is an unlikely cause of depopulation and abandonment of settlements at 1000 B.C. However, it is not unlikely that over the course of 400 years of settled village life, as growing populations became increasingly dependent upon drought-resistant barley agriculture to supplement hunting and foraging activities, unsustainable agricultural practices eventually resulted in local resource depletion, local ecological degradation, soil salinization, and socio-sanitation issues in Deccan Chalcolithic villages. In fact, there is support for this interpretation in the archaeological record (Kajale 1988). Large population centers like Daimabad did not adjust their behavior, suggesting perhaps that there was a lack of flexibility which may have contributed to their collapse. Small satellite settlements like Nevasa followed. Despite poverty, food shortages, poor sanitation, and declining health status, mid-sized settlements like Inamgaon retained enough flexibility to persist for some time. The archaeological record at Late Jorwe Inamgaon demonstrates that the population began to rely on a new mix of saline-tolerant crops, herding, increased foraging, and hunting activity. There were commensurate changes in the material culture and the meaning of structures built at the site as the population density slowly declined.

Reconstructing the demographic and childhood growth profiles can address this issue. In fact, it is not possible to evaluate whether the Late Jorwe was a time of degradation or prosperity without a demographic profile. Of course, paleodemography is challenging and the results are primarily useful as *relative* indicators of population dynamics in a comparative framework. Despite the challenges, paleodemography can be used to characterize the relationship between a population, its stress levels, and its environment within an adaptive framework. To do this, it is necessary to understand the range of normal for human communities. Fertility rates are principally influenced by the length of the reproductive period and the interbirth interval (IBI), the time between subsequent births. If we consider 15 years to be the minimum age of onset for the reproductive period in females, and the maximum age to be somewhere between 40 and 45 years, this would be equivalent to a reproductive period of 25–30 years. Often women do not reproduce for the maximum number of years theoretically possible—many will die during the reproductive period, not all women have access to full reproductive potential, fecundity (the capacity to become pregnant and successfully reproduce) varies, and control of fertility is common in non-industrialized and industrialized communities (Bogin 2001). A 15–25 year reproductive period is generally considered reasonable for human females in natural fertility populations (Bogin 2001; Livi-Bacci 2007).

The length of the IBI varies between human communities and is based on many biocultural circumstances, including duration of gestation and breast feeding, breast-feeding amennorhea, nutritional status, irregular ovulation cycles, cultural and familial traditions, and/or fetal mortality due to fetal wastage, miscarriage, abortion, and infanticide (Hrdy 1999; Gage 2000; Ellison 2001). When infant mortality is constant, the more offspring women bear by remaining in the reproductive cycle for longer and maintaining a short average IBI, the higher their individual fitness and reproductive success (Stevenson et al. 2004). However, IBI is biocultural, individual, and environmental (Pennington and Harpending 1988), and energetic trade-offs (Blurton-Jones and Sibley 1978) as well as different female strategies (Hrdy 1999) determine its length. Human IBI generally ranges from 1 to 5 years but often falls between 18 to 42 months (Livi-Bacci 2007). Historically, human populations generally maintained a GRR of 2–6 female offspring (4–12 total surviving offspring on average). But if females in a population had a long average reproductive span, near the maximum for humans (25 years), and a relatively short IBI of 18 months, they could accomplish a GRR of 8.35 (total fertility of 16.7 offspring) at the high end of the range of expectations for "normal" human fertility.

Based on my estimates for fertility and life expectancy at birth, it appears that Early Jorwe populations were heterogeneous. My results suggest that the large, fast-growing settlement at Daimabad had relatively lower fertility compared to all other Deccan Chalcolithic sites with a total fertility of 3.6 offspring. The estimated life expectancy at birth for this site was very long, indicating that the settlement growth was accomplished with relatively low reproductive costs for females. It is also likely that immigration into the village of Daimabad was partially responsible for the fast settlement growth rate in this village. However, in a relative sense, this profile suggests a low-pressure demographic situation with low reproductive costs and potentially lower stress levels for women and infants. Across the river, at the small, rural, satellite village of Nevasa, my estimates suggest the stable settlement growth rate in the Early Jorwe phase was maintained by moderately high fertility and a relatively shorter life expectancy at birth. These results support a suggestion made by Shinde (2002) that Nevasa represents a relatively poor, marginalized community on the periphery of the large regional center at Daimabad. The demographic profile does not contradict this idea, although additional research would be necessary to confirm that hypothesis.

High fertility also characterized the Early Jorwe phase at Inamgaon. For this village, a moderately fast settlement growth rate was associated with higher fertility and short life expectancy at birth compared to other Early Jorwe Chalcolithic sites. Thus the population at Inamgaon appears to have been relatively stressed, in the Early and the Late Jorwe. The evidence from the Late Jorwe at Inamgaon indicates that conditions did worsen still after 1000 B.C. In a time when many sites were abandoned, Inamgaon persisted but experienced a negative settlement growth rate accompanied by relatively high fertility and catastrophically low life expectancy at birth. This unsustainable high-pressure demographic situation would produce higher stress levels for women and infants in the Late Jorwe phase at Inamgaon and eventually may have contributed to the collapse of the settlement. However, given the high rate of fertility, the estimate for life expectancy at birth is likely underestimated to some extent, and declining settlement size also may have been related to increased rates of emigration away from the village.

The demographic profile also serves as a socio-cultural context in which to evaluate the biocultural stress markers. Biodemographic theory predicts a relationship between the growth of populations and the growth of individuals. I used estimates of body mass from the midshaft femur to compare emaciation among archaeological samples from the Deccan Chalcolithic. This approach has previously demonstrated that population differences in body mass index can be detected in archaeological populations with high biocultural stress levels (Robbins and Cowgill 2009). My results indicate that Late Jorwe Inamgaon (the population with the highest fertility rates and the shortest life expectancy at birth) also demonstrated higher frequency and greater severity of growth suppression in body mass index for young infants. In contrast, the Early Jorwe Inamgaon population demonstrated lower rates of growth suppression in body mass index for young infants. The results demonstrate that although the Early Jorwe population was stressed, mean BMI was significantly higher in this phase, actually matching contemporary reference standards from the Denver population. The reduction in mean BMI in the Late Jorwe supports an increase in stress levels in that skeletal sample.

How do these results compare with the results and interpretations of previous studies of growth disruption in enamel formation for Deccan Chalcolithic samples? Previous studies on enamel hypoplasia (Lukacs and Walimbe 1998, 2005; Lukacs 2007b) have demonstrated no significant difference in LEH for the permanent teeth, but the prevalence rate for LHPC in the primary dentition was significantly different. The Early Jorwe sample from Inamgaon had a higher frequency of LHPC compared to Nevasa, Daimabad, and the Late Jorwe. Thus, compared to all other Deccan Chalcolithic populations, the Early Jorwe Inamgaon population is characterized by high fertility, short life expectancy, lower frequency of acute stress (growth suppression in the skeleton), and greater frequency of chronic stress markers (LHPC). These results indicate that Early Jorwe infants who entered the death assemblage were "healthy survivors"—the lower mortality rate is associated with a lower proportion of acutely stressed, starving individuals and a higher proportion of individuals who lived long enough to express chronic stress. In contrast, the population from Late Jorwe Inamgaon had shorter life expectancy at birth, higher fertility, higher frequency of acute growth suppression in the skeleton, and lower frequency of chronic stress markers. The infants from this population appear to have been more likely to die from short-term stressors, succumbing to death before they could express chronic growth disruption.

Archaeologists and anthropologists have long sought to understand why Inamgaon was abandoned at the end of the Late Jorwe phase. The climate-culture change model attributes the culture changes to climate change, but new climate reconstructions do not support this idea. The subsistence transition model suggests that the Late Jorwe phase represented a successful adjustment to the decline of agriculture, with dietary diversity and lower settlement density leading to improvements in health, but this model never explained why Late Jorwe people also abandoned the settlement at Inamgaon after 300 years. New evidence from demography and skeletal growth profiles indicates that clearly this model is not entirely correct either.

My biodemography model offers another possible explanation for the end of the Deccan Chalcolithic. To briefly sum up, my research suggests that although agriculture is often associated with poor health in prehistory, the collapse of a major component of the subsistence system does not lead to improvements in health status. Instead, despite the attempted shift to reliance on other subsistence activities, the end of large-scale agricultural production at Inamgaon led to changes in the demographic dynamics, childhood health status, and eventually the abandonment of the settlement at Inamgaon. The growth of populations and the growth of individuals appear to have been closely related for Deccan Chalcolithic communities. Combining the two provides a powerful lens for understanding the changes that took place at the end of this period of Indian prehistory. Future work in this area should include reexamination of LHPC and its possible etiology, characterization of health in these samples using additional stress markers, and a consideration of demographic and biocultural responses to change in other communities occupied during this time in central India (e.g., Ramapuram).

This project confirms the importance of approaching the past using paleoclimatological, archaeological, and skeletal evidence to evaluate human populations and their response to environmental change. Through this approach, the Deccan Chalcolithic example demonstrates the importance of adaptability in determining the fate of communities facing local and/or global environmental change as historical, social, cultural, and ecological variation all play a role in the perception of the environment, acceptability of risk, and the adjustments that communities conceive and implement. Additional archaeological research that takes advantage of recent paleoclimate research and asks questions about local cognitive and behavioral strategies for coping with environmental change will undoubtedly result in a more complex picture of prehistoric populations in India.

Appendix A

Burials from Daimabad: Archaeological Context and Grave Goods

GR #	Trench	Layer	Period	Grave goods	Type
1	CZ–FZ 52–61	S1 (2)	J	I d, 1 st	Aiva
2	CZ61	(2)	J	1 mst	Aivb
3	CZ60	(2)	J	1 bc	Aivc
4	DZ55	(1A)	J	1 c, 1 sb	V
5	.2m E 4	(1A)	J		V
6 fd	DZ60	S2, S3	J		A
7* f	CZ59 #8		J	3 c, 1 ob, f	A
8* f	CZ–DZ 59	S2 (4)	J	3 c	A
9	DZ59	S2A (3–4)	J	1 c	Aiii
10	CZ60	S3 (4)	J		Ai
11	CZ60 center	S3 (4)	J	1 c, 1 ab	Ai
12	CZ60 1mE11	S3 (4)	J	1 cb	Ai
13	CZ60 .5mE12	S3 (4)	J	Sb	Ai
14	DZ58 near 15	S2 (2A-3)	J	Cf	Aiii
15 ne	DZ58	S2 (2A-3)	J		Ai
16 nr	DZ58	S2A (3)	J		Ai
17	EZ55		J		C
18	CZ61		LH	Mb	Ext
19 d	L48 E section	S2 (3–5)	J		Av
20	X4	S7 (8–11)	M	72 sb	Ci
21	Y3		M	2 c	Ai
22 s, t	Y3	S7 (8–9)	M		Cii
23	Y3?	S7 (8–9)	M	4 c, st	Ciii
24	X3	S3 (4–5)	M	1 b	Ai
25	Z3	S3 (4–5)	M		Biii
26 s, t	X3	S6 (7–8)	M		D
27 s	X5	S5 (6–8)	M	2 c	Ai
28	X5		M		Ai
29	X5	S5 (6–7)	M		Bii
30 ne	FZ64–65 wall	S4 (5–6)	J		Ai
31 ne	FZ64–65 wall	S4 (5–6)	J		Ai
32 ne	FZ64–65 wall	S4 (5–6)	J	1 c	Ai
33	FZ63	S12 (13–15)	D	i u, 3 c	Aii
34 a, s	Z4, II	S9 (blk soil)	D	5 c	B
35* d	Z1	S1 (2–3)	J	1 c, ch bl	Ai

continued

GR #	Trench	Layer	Period	Grave goods	Type
36			J		Ai
37	BZ4 wall	S1	J	2 c	Aiia
38 su	BZ3	S1	J	2 c	Ai
39	BZ4 wall	S1	J	2 c	Aiia
40 su	BZ4	S1	J	9 c	Ai
41	AZ 3–4	S3	J		Ai
42	FZ3–BZ4 wall	S1Near 37	J	up, 1c	Aiib
43	BZ2		J	3 c, 2 cb, 4 sb	Ai
44	AZ3	S4	M-J		Aii
45	AZ3 E wall	S4	M-J		Ai
46	BZ2 N wall	S2	J		Aiia
47	BZ3	S4 (5)	M-J		Ai
48	BZ1	S1 (2–4)	J		Aiii
49	BZ1	S1B (2)	J		Aiii
50	.5m e of AST	S4	M-J		Ai
51	.5m e of #50	S4 (5)	M-J		Ai
52	CZ3	S1B (3–4)	J	1c	Aiia
53 nf	CZ3		J	p, 2 c, 1 s, 1 st	E pit
54	ZD62	S1 (2–4)	J		C
55 d	ZD62	S5 (6–8)	M		Bi
56	ZD61	S5 (6–8)	M		Aiia
57	BZ3	S4 (5–6)	M-J	pd	Bi
58 nr	BZ3 1mE 57	(6)	M-J	pd	Bii
59 c, a	ZD61	S7 (8–10)	D		C
60 nr	Z69	S5 (6–8)	J		Ai
61	Z69 0.3mE 60	S (7–8)	J	3 c	Ai
62 nr	Z–AZ 69–70	S12	M		Ai
63 pr, s	Z–AZ 69–70	S14	M	ad	Biv
64 nr	Z–AZ 69–70	S14	M		Ai
65 su			J		Ai
66 su			J	Smallest urn	Ai
67	~ cut into 66		J	2 c	Ai
68	DZ4	Pit w 69 70	J	1 c	Avi
69 su	DZ4	Pit w 68 70	J		Avi
70 su	DZ4	Pit w 68 69	J		Avi
71	DZ4 W 68–70	S1	J		C
72			J	1 cg	Ai
73 su	EZ4 1mW 72		J		Ai
74	EZ4 .4mW73		J	2 c	Aiia
75	ZD60	S (6–7) II	M	pd, 255 sb, 22 cb, 1 c	Aiib

Notes: GR number numbering system was created for the burials to simplify the labeling of individuals in the collection.

* = skull bones within concave-sided carinated bowl; a = ash within burial or urn; c = bones appeared charred; d = excavator reported damage to burial; f = excavator reported that feet appeared to have been chopped off; fd = burial disturbed by flood (layer 2 a flood born deposit); ne = burial not fully excavated;

nf = excavator reported that feet did not appear to have been chopped off; nr = burial urns not removed; s = symbolic; su = small-sized urn; t = burial appeared to excavator to have been lined with fibrous twigs; S# = undisturbed sealing layer overlying the burial.

LH = Late Harappan; Daim = Daimabad; M = Malwa; M-J Transition between Malwa and Jorwe; J = Jorwe; c = burial in house courtyard; f = burial was cut into house floor; ad = urn had appliqué design; b = beads (ab = agate, cb = carnelian, ob = onyx, sb = shell); bc = basalt chopper; c = ceramics; cf = burial pit filled with hard clay; ch = bl chalcedony blade; cg = copper goddess figurine; f = flowers; id = urn had incised designs; iu = urn had impressed designs; mb = mud-brick lined grave; mst = muller stone; p = plastered grave; pd = urn had painted designs; pr = purple-red urn; s = fresh water shell; st = stone; up = urn was painted.

Double urn burials (Type A): Ai = mouth to mouth urns; Aii = mouth in mouth; Aiia = burnished gray ware with the northern urn inside the mouth of the southern; Aiib = with the northern urn of burnished gray ware placed up to the neck of the vessel within the southern urn of Malwa ware and both sealed with a clay matrix.

Single urn burials (Type B): Bi = single urn of Malwa ware positioned vertically, the mouth covered with a shallow bowl of burnished gray ware in an inverted position; Bii = small burnished gray ware vases with a squat body and grooved neck, also positioned vertically, and covered with a lid of the same ceramic body; Biii = bowl of gray ware placed horizontally within the mouth of the urn; Biv = symbolic burials in urn of purplish redware placed vertically.

Burial pit with associated ceramics (Type C): Ci = three bowls of Malwa ware, two placed mouth to mouth in a north-south orientation and the third placed vertically; Cii = three bowls of Malwa ware placed in a row with their mouths facing south and fibrous twigs lining pit; Ciii = with bowls of black burnished ware, stones placed above the burial and fibrous twigs lined the base.

Pit Burials without associated pottery (Type D): Ext = extended burial.

Appendix B

Age Estimates for Subadults in Chalcolithic Samples

Age estimation for perinates ≤ 44 lunar weeks at Inamgaon (ages in lunar weeks).

ind	Phase	Age range	Age	Petrosa	Basilaris	Lateralis	Sphenoid	Malar	Clavicle	Scapula	Ilium	Ischium	Pubis	Long bones
47	EJ	28–32	31.2	30	30						32			30–32
74	EJ	34–38	37.1	34–36	36–38		perinatal					32	34–36	36–38
97	EJ	36–40	37.0											36–38
103a	EJ	36–40	37.0	36			38–40							36–38
78	EJ	36–44	39.1	26										38–40
124	EJ	36–44	39.7											38–40
80	EJ	36–44	40.5	32										39–41
84	EJ	36–44	40.8								38			39–41
103b	EJ	36–44	40.5	38					38					39–41
83	EJ	40–44	42.0	0–3			-1	-0.5						40–44
142	LJ	28–32	29.8											28–32
215	LJ	36–44	37.3											38–40
144	LJ	36–40	38.3			38		26–28	38					38–40
1	LJ	38–40	39.7								40			39–41
18	LJ	36–44	38.5	32–34										38–40
118	LJ	36–44	40.0			40			38		38–40			39–41
5	LJ	36–44	40											perinatal
22	LJ	36–44	40											perinatal
69	LJ	36–44	40.8	28–30		perinatal	40+	36	40		40			
150	LJ	36–44	40					34–36						
166	LJ	36–44	40	32–34					perinatal					
239	LJ	36–44	41.0	32–34										39–41

continued

ind	Phase	Age range	Age	Dentition	Mand sym	Frontal sym	Temporal	Basilaris	Lateralis	Sphenoid	Malar	Clavicle, scapula	Pelvic bones	Long bones
170	LJ	36–44	39.7				40	40	28–34	40		40+		39–41
207	LJ	36–44	40											perinatal
231	LJ	36–44	40			perinatal								
234	LJ	36–44	40	perinatal					perinatal					
63b	LJ	40–44	41.4	0–3			-0.5	-0.5	0	perinatal		-0.5		-0.5
155	LJ	40–44	41.9	2–5										>0
63a	LJ	40–44	42.8			30	30			30	30	-2.5	0+	40–44
141	M	28–32	30	30		36	36	32		36	36–38	30	28	30–32
94	M	34–38	36	34–36			36	32		36		36–38		34–36
110	M	32–40	36											32–36
137	M	36–40	38	38–40		38	38	38–40	38	38–40	38	38		38–40
130a	M	36–40	38				36		30–32					36–38
130b	M	36–40	38								38	perinatal		34–38
112	M	36–44	40							36–38	38	36–38		38–40
135	M	36–44	40	40		40	40	40	36–38	40	38–40	38–40	40	38–40
139	M	36–44	40	40		36	36			38	38	38		38–40
129a	M	36–44	40	32–40		40	40	40	40	<6mo	36–40	40		37–42
129b1	M	36–44	40			perinatal								38–40

Age estimation for infants 1–12 months at Inamgaon (ages in months).

ind	Phase	Age range	Age	Dentition	Mand sym	Frontal sym	Temporal	Basilaris	Lateralis	Sphenoid	Malar	Clavicle, scapula	Pelvic bones	Long bones
88	EJ	1–3	2											0–6
102	EJ	1–3	2										0+	0–6
61	EJ	2–4	3	2–4				3	0–12	0+			0+	0–6
87	EJ	1–5	3	1–5										0–6

ID	Type										
114a	EJ	2–4	3								3–6
48	EJ	2–6	4								2–6
95	EJ	3–5	4	3–5							0–6
120	EJ	3–5	4	3–5							
81	EJ	5–7	6	5–7			5				6–12
105	EJ	4–8	6	4–8							3–9
86	EJ	6–12	9			12	5–12	<24			6–12
99	EJ	6–12	9								
111	EJ	6–12	9							7–12	6–12
122	EJ	6–12	9	6–9	12		9				6–12
79	EJ	9–13	11	9–12	12						0–3
7	LJ	0–2	1								
225	LJ	0–2	1								0–3
180	LJ	0–4	2								0–6
226	LJ	1–3	2	0–3							0–6
179b	LJ	1–3	2	0–3							0–6
194b	LJ	1–3	2	0–3			3		0–6		0–6
16	LJ	2–4	3	2–4			3				0–6
151	LJ	2–4	3	2–4							
174	LJ	2–4	3	2–4							
212	LJ	2–4	3	0–4		0+				−0.5	
233	LJ	0–6	3								0–6
205a	LJ	2–4	3	2–4							0–6
4	LJ	3–5	4	3–5							0–6
96	LJ	3–5	4	3–5		0–12	3			5–7	3–6
92	LJ	4–6	5	4–6							

continued

ind	Phase	Age range	Age	Dentition	Mand sym	Frontal sym	Temporal	Basilaris	Lateralis	Sphenoid	Malar	Clavicle, scapula	Pelvic bones	Long bones
20	LJ	5–7	6	5–7										
55	LJ	5–7	6	5–7										
147	LJ	0–12	6											0–12
149	LJ	3–9	6	3–9										
203	LJ	0–12	6											0–12
222	LJ	5–7	6	5–7				6						
235	LJ	4–8	6	3–9										0–12
238	LJ	0–12	6	3–12										0–12
17	LJ	4–9	7	4–9				<8						
98	LJ	6–8	7				0–12	5–8						3–9
3	LJ	6–12	9	6–12					<12					6–12
148	LJ	6–12	9	6–12			12	13						6–12
158	LJ	6–12	9	6–12										6–12
162	LJ	6–12	9	6–12						12				6–12
169	LJ	6–12	9	6–12		12–24	9–12	8–11						6–12
186	LJ	6–12	9	6–12				11	9–12			7–12		6–12
211	LJ	6–12	9	6–12	12		12							
213	LJ	6–12	9	6–12	12									3–9
223	LJ	6–12	9	6–12				8	<12			12		6–12
247	LJ	6–12	9	6–12										
179a	LJ	6–12	9			12–24								6–12
205b	LJ	6–12	9			<12								6–12
201	LJ	9–13	11		12			12						6–12

ind	Phase	Age range	Age	Dentition	Mand symph	Frontal symph	Temporal	Basilaris	Lateralis	Malar	Vertebrae	Clavicle, scapula	Pelvic bones	Long bones
138	M	1-3	2	0-3	12		12	0+	0+	12	0+		0+	0-6
127	M	6-8	7	4-9										3-9
129b	M	6-12	9	0-3					6-12					6-12
136	M	9-15	12	9-15	12				12					0-6
120	?	3-5	4	3-6	12					12				0-6

Age estimation for subadults 13-60 months at Inamgaon (ages in months).

ind	Phase	Age range	Age	Dentition	Mand symph	Frontal symph	Temporal	Basilaris	Lateralis	Malar	Vertebrae	Clavicle, scapula	Pelvic bones	Long bones
105a	EJ	9-15	13	9-15										6-12
114	EJ	6-12	13	6-12			12	9	12	12				6-12
123	EJ	12-18	15	12-18	12			13					12	6-12
73	EJ	18-24	21	18-24			>12							18-24
62	EJ	30-42	36	30-42										
85	EJ	30-42	36	30-42										36-48
100	EJ	30-42	36	30-42										
124w	EJ	24-48	36	30-42		24-48	>12		24-48		36-48	18-24		24-36
12	EJ	36-48	42	36-48					24-48					36-48
89	EJ	36-48	42	36-48		24-48	>12							36-48
115	EJ	36-48	42								36-48			
119	EJ	36-48	42	36-48	>12	24-48	>12	14	24-48	24-36	<36	24-36		24
126	EJ	36-48	42	36-42										
198	EJ	48-72	60	54-66							48-60			42-54
6	LJ	9-15	13	12-14										6-12
58	LJ	9-15	13	12-14	>12	<24	6-12	12-36						6-12

continued

ind	Phase	Age range	Age	Dentition	Mand symph	Frontal symph	Temporal	Basilaris	Lateralis	Malar	Vertebrae	Clavicle, scapula	Pelvic bones	Long bones
93	LJ	9–15	13	12–14		<24						7–12		
154	LJ	9–15	13	12–18	>12									9–15
156	LJ	9–15	13	12–14	>12		12	13		12–24				9–15
171	LJ	9–15	13	9–15	>12	<12	12							9–15
187	LJ	9–15	13	12–14	>12	12	12	18						9–15
204	LJ	9–15	13											9–15
143	LJ	12–18	15	12–18										
188	LJ	12–18	15	12–18	>12	24			12		12			
9	LJ	18–24	21	18–24		24	12							
3a	LJ	18–24	21	18–24										
145	LJ	24–36	30	30–36										24–36
143a	LJ	30–36	33	30–36										
117	LJ	30–42	36	30–42		24+	12+	27	24–48			12–24		
176	LJ	30–42	36	30–42		24–48	>12	15–18	>24					18–30
177	LJ	30–42	36	30–42										
237	LJ	24–48	36	36–48				27			<36–48			
245	LJ	30–42	36	30–42							36–48			24–36
206c	LJ	24–48	36	30–42	>12									
228a	LJ	24–48	36	30–36	>12	24–48	>12					24–36		24–36
240a	LJ	24–48	36	30–42										
163	LJ	36–48	42	36–48		24–48					36–48			24–36
167	LJ	36–48	42	36–48							36–48			
230	LJ	36–48	42	36–48							<36–48			
228b	LJ	36–48	42											36–48
H105	LJ	42–54	48	42–54			<60				<48–60			24–36

ind	Phase	Age range	Age	Dentition	Vertebrae	Clavicle	Phalanges	Ischium	Pubis	Long bones	Epiphyseal centers
181	LJ	48-60	54	48-60							24-36
185	LJ	48-60	54	48-60							36-48
194a	LJ	48-60	54	48-60					48-72	36-48	
32	LJ	48-72	60				<60-84				
216	LJ	54-66	60	54-66			60-84				42-54
133	M	18-30	24	18-30	>12						

Age Estimation for older children and adolescents 61–156 months at Inamgaon (ages in months).

ind	Phase	Age range	Age	Dentition	Vertebrae	Clavicle	Phalanges	Ischium	Pubis	Long bones	Epiphyseal centers
124e	EJ	66-78	72	66-78	66-78					48-60	
72	EJ	72-96	84	72-96							
124c	EJ	84-96	90	84-96						60-72	
101	EJ	96-108	102							72-84	>96
68	EJ	144-168	156	>144	>144					12-13	12-14
182	LJ	60-72	66	60-72	<72						
221a	LJ	60-72	66	60-72		60-72				54-66	
52	LJ	66-78	72	66-78					60-72	60-72	72
183	LJ	60-84	72	66-78	>60			<72		54-66	
196	LJ	60-84	72	66-78	60-72	72		<60-108		72-84	>60
175a	LJ	60-84	72	72-92							72-84
184	LJ	72-96	84	72-92	<72					72-84	
192	LJ	72-96	84	78-92	>72	72				60-72	
191	LJ	84-96	90	84-96						84-96	
250	LJ	84-96	90	84-96						84-96	
175b	LJ	84-96	90	84-96	<72					60-72	<136

continued

GR	Ind	Age range	Age							
208a	LJ	84–96	90	84–96	>72				72–84	72–84
221b	LJ	84–96	90	84–96	>72				72–84	72–84
240	LJ	108–120	114	108–120	96–108	<96			72–84	72–84
241	LJ	108–120	114	108–120						
15	LJ	120–144	132	120–144						
65	LJ	120–144	132	120–144				<168		>120
195	LJ	120–144	132	120–144				<168		
190	LJ	144–168	144				<156			
192a	LJ	132–156	144	144	144					>120
160	LJ	144–168	156	144–168						
197	LJ	144–168	156	144–168	>144				144–168	
228	LJ	144–168	156	144–168	>144	>120		<168	144–168	
189	LJ	144–168	156	144–168					<180	
206e	LJ	168–192	180					<204	168–192	
176a	LJ	168–192	180	168–192						
199a	LJ	168–192	180	144–228				168–204	168–192	
168	LJ	180–196	192	180–196					180–196	
218	LJ	180–228	204	180–196					180–204	

Age estimation for perinates ≤ 44 lunar weeks at Daimabad (ages in lunar weeks).

GR	Ind	Age range	Age	Basilaris	Sphenoid	Clavicle	Long bones
30	35(69)	35–37	36.3				35–37
34	39(73)	35–37	36.8	38	38–39		35–37
28	33(67)	36–38	37.1			39	36–38
14	16(50)	36–39	37.7				36–39
11	13(47)	36–39	37.8				36–39

33	38(72)	37.8	36–39
10	11(45)	39.4	38–40
35	40(74)	39.5	38–41
15	17 (51)	39.9	38–42
16	17 (51b)	40.8	39–42
7	8(42)	38	36–40
26	31(65)	39	38–42
21	22(56)	40	38–42

Age estimation for infants 1-12 months at Daimabad (ages in months).

GR #	Ind	Age range	Age	Dentition	Mand sym	Frontal sym	Temporal	Lateralis	Sphenoid	Vertebrae	Clavicle	Ilium	Long bones
13	15(49)	1–3	2									>1	1–3
31	36(70)	1–3	2										1–3
4	5(39)	1–6	3				1–3	<6					
36	41(75)	1–6	3	2–4									1–3
9	10 (44) teeth	1–6	3	1–6									
9	10 (44) skl	< 3	3										
3	4(38)	4–6	5	4–6			5–6						3–6
32	37 (71)	1–6	3										1–6
12	14(48)	8–12	10	8–12									
2	3 (37)	6–12	9	6–12									

continued

Age estimation for subadults 12–60 months at Daimabad (ages in months).

GR #	Ind.	Age range	Age	Dentition	Frontal symph	Long bones
1	2 (36)	10–14	13	12–14		
8	9(43)	9–15	13			9–15
29	34 (68)	9–15	13	12–14	>12	9–15
25	30(64)	12–15	14	12–15		12–18
27	32 (66)	12–15	14	12–15		9–12
5	7 (41)	12–18	15			12–18
24	28(62)	15–21	18		24	18
22	26(60)	12–24	18			18
18	20a	30–42	36	30–42		
6	8	36–42	39	36–42		36–48
19	20b	36–48	42	36–48		
20	20c	36–48	42	36–48		
23	27 (61)	42–54	48	42–54		

Age estimation for perinates < 44 lunar weeks at Nevasa (ages in lunar weeks).

VM	Age range	Age	Dentition	Petrosa	Sphenoid	Ilium	Long bones
5	38–42	40					perinatal
9	38–42	40					36–44
11	38–42	40					36–44
10	38–42	40	36–44			40	36–44
15	38–42	40					36–44
7	38.4	42.0					37–39
17	38.4	38.4					37–39
23	38.8	38.8	36–44	32	< 40	40	38–40
6	39.7	39.7					38–40
33	40–43	41				perinatal	

Age estimation for infants 1–12 months at Nevasa (ages in months).

VM	Age range	Age	Dentition	Mand sym	Frontal sym	Temporal	Basilaris	Lateralis	Sphenoid	Vertebrae	Clavicle	Long bones
24	42–46	1										
60	42–46	1										
12	1–5	3										2–4
16	0–6	3										2–4
32	0–6	3										
35	2–4	3										1–3
57	3–6	5	3–6									3–6
18	0–12	6										
34	0–12	6										
36	3–9	6	3–9				5			>12	24	3–6
50	3–9	6	3–9		< 24	>12		< 12				3–6
53	3–9	6	3–9	>12		>12	5	< 24		>12		3–9
54	3–9	6	3–9	>12		>12	5	< 12				3–6
14	6–9	8	6–9									
1	6–12	9	6–12	>12	>24	>12			>12	< 12	7–12	
20	6–12	9	6–12			>12						
25	6–12	9					14					6–9
27	6–12	9										6–9
67	6–12	9	6–12									
31	9–12	11	9–12		< 24	>12			>12			6–9
2	9–15	12	9–15		>24	>12	12–18	< 12	>12	>12		6–12

continued

Age estimation for subadults 13–60 months at Nevasa (ages in months).

VM	Age range	Age	Dentition	Mand symph	Frontal symph	Temporal	Basilaris	Lateralis	Sphenoid	Vertebrae	Scapula	Long bones	Epiphyseal centers
4	9–15	13	12–15	>12	<24		>12	>12					
8a	9–15	13										9–15	
13	9–15	13								>12		9–15	
29	9–15	13								<36		9–15	
37	9–15	13	12–15		<24							9–15	
45	9–15	13											
49	9–15	13	12–15		>24	>12		<24					
59	9–15	13										9–15	
61	9–15	13	12–15	>12									
70	9–15	13										9–15	
21	12–18	15	12–18										
28	12–18	15	12–18		>24	>12			>12			9–15	
26	12–18	15	12–18										
38	12–18	15	12–18		>24		12–16			>12	6–12	12–18	
44	12–18	15	12–18	>12									
55	12–18	15	12–18		<24			<12					
52	12–24	18	15–21		>24	>12				>12		6–9	
63	12–18	15	12–18	>12									
64	12–18	15	12–18										
66	12–24	18	15–21										
42	18–24	21	18–24		>24								
46	21–27	24	21–24					<24		12–36		18–24	
3	24–36	30	24–36	>24	>12			>12					

VM													
41	24–36	30	24–36	>12	>24		>12	21–27	>24		12–36	18–24	>24
65	24–36	30	24–36										
40	24–36	30	24–36	>12									
68	30–42	36	30–42	>12									
8	24–48	36										30–42	
51	30–42	36	30–42	>12									
62	30–42	36	30–42										

Age estimation for subadults 13–60 months at Nevasa, 9-15.

VM	Age range	Age	Dentition	Mand symph	Frontal symph	Temporal Basilaris	Lateralis	Sphenoid	Vertebrae	Scapula	Long bones	Epiphyseal centers
22	30–42	36	30–42									
39	36–48	42	36–48						< 36–48			24–60
43	36–48	42	36–48						36–48			24–60
56	36–48	42	36–48						< 36–48			
58a	24–72	48	36–60									
48	54–66	60	54–66									

Age estimation for subadults 61–156 months from Nevasa (ages in months).

VM	Age range	Age	Dentition	Vertebrae	Epiphyseal centers
58	60–84	72	60–84		
30	108–132	120	108–132		
47	108–132	120	108–132	120–144	~120
69	108–132	120	108–132		~120

Appendix C

Long Bone Lengths (mm) and Stature (cm) for Individuals with Dental Age Estimates (months)

Early Jorwe Inamgaon

ind	Age category	Stature	Humerus	Radius	Ulna	Femur	Tibia	Fibula
83	2	53.8			65.0		70.0*	
61	3	52.0		50.0				
87	3	68.9		62.5*	69.0			
95	4	71.0		65.0				
105	6	71.1		65.0*	80.0			
81	6	74.0			85.0			92.0*
122	9	87.6	108.0		83.0	110.0*		
114a	9	66.0		67.0		105.0*		
123	15	74.7			90.0			
73	21	78.7	115.0			140.0*		
85	36	98.3		118.0*	113.0			
124w	36	88.5			113.0		155.0*	
119	42	85.77	131.0					
89	42	94.2	150.0	115.0			170.0*	155.0
12	42	98.0		120.0				
198	60	104.5		127.0	140.0*	220.0	184.0	

Late Jorwe Inamgaon

ind	Age category	Stature	Humerus	Radius	Ulna	Femur	Tibia	Fibula
18	0	50.4	64.1*					
170	0	52.8	67.0	52.0		77.1*	68.0	
155	2	45.0			65.0*			
63b	2	57.9		56.6*				
226	2	62.7	70.0*					
194b	2	63.5	71.0*					
16	3	70.0	80.0*					

continued

VM[a]	Age	Stature	Humerus	Radius	Ulna	Femur	Tibia	Fibula
96	4	70.5	79.3*	66.8	75.7			
4	4	85.0				100.0*		80.0
162	9	76.0	87.0*					
3	9	78.0				120.0*		
186	9	78.4	90.0*					
148	9	78.7	100.0	76.0		124.0*		
213	9	80.0	86.0*	77.0				
223	9	81.6				130.0*		
154	12	73.0			87.0*			
171	12	82.8		76.0*			100.0	
6	12	87.0			80.0*			
58	12	87.0	98.0*	80.0	88.0			
187	12	87.2		80.2*	90.8			
156	12	88.1			88.0*			
176	36	81.8	122.0	98.0	110.0	160.0*	136.0	133.0
228a	36	86.2	130.0*					
245	36	86.7	133.0*	110.0	127.0			
163	42	84.9	129.0				150.0*	
h 105	48	89.6			115.0*			
185	54	93.5	135.0			195.0*		159.0
194a	54	96.4	148.0		132.0	215.0*		175.0
216	60	98.1	159.0			213.0*		

Nevasa

VM[a]	Age	Stature	Humerus	Radius	Ulna	Femur	Tibia	Fibula
23	0	50.4	64.0	51.3		76.0*	64.4	
57	5	69.0	78.0*	61.9	68.2			
54	6	67.7	76.3	62.4		89.0*		
36	6	76.1	87.0		75.3	95.0*	88.0	80.0
53	6	70.2				95.0*	86.0	
2	12	68.6	92.0			125.0*		
28	15	82.4	95.0*					
21	15	80.6			85.0*			
38	15	84.7	104.0		89.5		107.2*	
52	18	71.5		76.0*				
46	24	81.5			110.0*			124.7
41	30	80.9		90.7*				
8	36	89.7	140.0*					

continued

Daimabad

GR[b]	Age	Stature	Humerus	Radius	Ulna	Femur	Tibia	Fibula
3	5	64.0	75.0				85.0*	
29	12	76.4		76.8		105.0*		
25	15	80.7			100.0*			
27	15	80.4	95.0			120.0*		105.0

*Bone used for stature estimate.
[a]Individuals listed using VM numbers for Nevasa (Mushrif and Walimbe 2006).
[b]Individuals listed using GR numbers for Daimabad (see Appendix A).

Appendix D

Midshaft Femur Cross-Section Measurements for All Individuals with Intact Femur Midshafts

Population	Individual	Age category	Transverse diameter	Medial cortex	Lateral cortex	J	Zp
EJ INM	74	0	5.7	2.0	2.6	101	35.7
EJ INM	97	0	6.4	1.9	1.3	152	47.8
EJ INM	103a	0	5.4	1.9	2.1	85	31.3
EJ INM	80	0	5.4	2.2	1.8	81	30.3
EJ INM	61	0	7.4	1.5	1.4	254	68.6
EJ INM	48	0	9.7	1.8	1.8	727	150.4
EJ INM	105	1	7.4	1.3	1.7	260	70.2
EJ INM	111	1	10.9	1.6	1.7	1051	192.8
EJ INM	122	1	9.5	2.6	2.1	736	155.5
EJ INM	100	3	11.6	2.2	2.1	1483	255.9
EJ INM	119	3	18.8	4.1	3.8	1924	261.5
EJ INM	12	5	14.6	3.0	2.6	3769	517.7
LJ INM	142	0	5.3	1.8	2.0	76	28.7
LJ INM	144	0	5.1	1.4	1.3	64	25.1
LJ INM	1	0	6.1	2.2	2.2	137	44.8
LJ INM	22	0	6.7	1.5	1.6	177	53.1
LJ INM	170S	0	6.0	1.4	1.7	123	40.9
LJ INM	179b	0	7.7	2.9	2.7	338	88.2
LJ INM	212–1	0	8.7	2.6	2.7	558	127.8
LJ INM	212–2	0	10.5	3.2	2.6	1147	218.3
LJ INM	4	0	9.5	1.8	2.7	731	154.4
LJ INM	147	1	9.1	1.7	1.8	568	125.6
LJ INM	3	1	10.3	1.5	2.0	896	174.2
LJ INM	148	1	9.8	2.6	2.4	844	172.7
LJ INM	169S	1	10.9	2.1	2.3	1221	223.4
LJ INM	213	1	9.1	.9	1.5	475	104.4
LJ INM	223	1	8.3	2.0	1.3	409	98.3
LJ INM	162	1	10.0	1.7	2.6	864	173.7
LJ INM	179a	1	10.2	2.7	3.5	1041	203.9
LJ INM	171	1	10.9	2.0	2.1	1177	215.9
LJ INM	187	1	9.5	1.5	1.8	641	135.5
LJ INM	145	3	12.2	2.0	2.1	1735	285.4

continued

Population	Individual	Age category	Transverse diameter	Medial cortex	Lateral cortex	J	Zp
LJ INM	176	3	13.7	2.6	2.6	2910	425.4
LJ INM	206c	3	12.7	2.8	2.6	2291	359.9
LJ INM	228a	3	11.9	2.8	2.6	1810	303.3
LJ INM	230	4	13.9	3.0	2.2	3072	442.3
LJ INM	194a	5	14.3	2.4	2.7	3372	472.6
LJ INM	216	5	13.5	2.6	2.8	2824	419.0
LJ INM	182a	5	12.8	3.0	3.3	2439	381.9
LJ INM	221a	5	14.8	2.7	3.0	4072	549.6
NVS	23	0	6.3	2.1	2.2	156	49.3
NVS	25	1	9.9	2.8	3.3	911	184.6
NVS	28	1	10.2	2.6	2.2	975	191.3
NVS	38	1	11.7	2.1	2.1	1525	260.7
NVS	55	1	8.2	2.0	1.5	398	97.0
NVS	52	2	10.4	1.5	1.7	872	167.9
NVS	46	2	12.6	2.8	3.3	2285	363.6
NVS	3	3	10.3	1.3	2.3	906	175.8
NVS	41	3	10.9	3.2	2.6	1326	242.8
NVS	39	4	13.0	4.1	3.1	2655	410.0
DMD	13	0	6.7	2.0	2.9	199	59.3
DMD	16	0	6.1	2.0	2.4	138	44.9
DMD	17	0	6.0	1.6	1.4	116	38.9
DMD	38	0	6.4	1.4	2.1	153	48.3
DMD	11	0	6.5	2.5	2.7	177	54.3
DMD	35	0	5.5	1.8	2.3	92	33.2
DMD	8	0	7.5	1.8	1.9	292	77.7
DMD	15	0	9.0	2.5	2.8	615	137.3
DMD	32	1	9.9	1.7	1.8	792	159.4

Appendix E

Stature and Body Mass Estimates for Deccan Chalcolithic Specimens

Population	Individual	Age category	Bone used for stature	Stature estimate (cm)	J	Body mass estimate (kg)
EJ INM	74	0	Femur[a]	48.5	101	3.9
EJ INM	97	0	Humerus	54.9	152	4.2
EJ INM	103a	0	Humerus	47.2	85	3.8
EJ INM	80	0	Fibula	50.0	81	3.8
EJ INM	61	0	Radius	52.0	254	4.8
EJ INM	105	1	Radius[b]	71.1	260	7.3
EJ INM	111	1	Femur	71.8	1051	8.9
EJ INM	122	1	Femur	65.9	736	8.3
LJ INM	142	0	Humerus	36.7	76	3.8
LJ INM	144	0	Femur	50.5	64	3.7
LJ INM	1	0	Femur	51.6	137	4.1
LJ INM	22	0	Fibula	57.7	177	4.4
LJ INM	170S	0	Humerus	52.8	123	4.0
LJ INM	4	0	Femur	62.9	731	7.7
LJ INM	3	1	Femur	69.0	896	8.6
LJ INM	148	1	Femur	70.2	844	8.5
LJ INM	169S	1	Fibula[c]	75.9	1221	9.2
LJ INM	213	1	Radius	80.0	475	7.7
LJ INM	223	1	Femur	72.0	409	7.6
LJ INM	162	1	Humerus	76.0	864	8.5
LJ INM	179a	1	Tibia	78.8	1041	8.9
LJ INM	171	1	Radius	82.8	1177	9.1
LJ INM	187	1	Radius	87.2	641	8.1
LJ INM	145	3	Femur	88.0	1735	12.3
LJ INM	176	3	Femur	160.0	2910	13.4
LJ INM	194a	5	Humerus	101.0	3372	16.1
LJ INM	216	5	Femur	100.3	2824	15.6
NVS	25	1	Tibia	73.8	911	8.6
NVS	28	1	Humerus	82.4	975	8.7
NVS	38	1	Humerus	84.7	1525	9.8
NVS	52	2	Radius	71.5	872	9.9
NVS	46	2	Fibula[c]	81.5	2285	12.7
DMD	13	0	Femur	64.0	199	4.5
DMD	16	0	Tibia	54.0	138	4.1

continued

Population	Individual	Age category	Bone used for stature	Stature estimate (cm)	J	Body mass estimate (kg)
DMD	11	0	Femur	50.7	177	4.4
DMD	35	0	Ulna	52.2	92	3.9
DMD	8	0	Humerus	83.3	292	5.1
DMD	15	0	Femur	56.3	615	7.0

[a]Stature estimates for perinates (< 44 lunar weeks) were made from equations provided by Fazekas and Kosa (1978).

[b]Stature estimates from the humerus, radius, femur, or tibia of individuals 6 months and older were made using formulas in Ruff (2007).

[c]Stature estimates for the fibula in individuals 6 months and older were made using Maresh (1970).

Notes

Chapter 1. Origins

1. This estimate is based on the calculation that there were 200 people per square hectare at Chalcolithic settlements (Dhavalikar 1988b).

2. Demography is the study of population statistics, in this case total fertility rates (number of offspring per woman, given she conforms to age specific mortality rates), infant mortality (number of offspring dying in the first year of life), life expectancy at birth (average number of years one is expected to live at the moment of birth), and population growth rates (crude birth rate minus crude death rate plus immigration rate). In paleodemography these variables are estimates. The total population size and the immigration rate are unknown.

3. Panja's model (1996, 1999, 2003) is a modification of Dhavalikar's model, which suggests the Late Jorwe at Inamgaon was a time when much of the village was abandoned. Evidence from site-formation processes and faunal analysis was interpreted as indicating the site was abandoned and reoccupied by hunting, foraging, and herding people only seasonally during the Late Jorwe. While this model is considered here in the sense that I have incorporated the idea that much of the site was abandoned, a test of whether the same people occupied the site is beyond the scope of this book. However, the ceramic traditions and other artifacts, as well as the burial practices, suggest continuity.

Chapter 2. The Western Deccan Plateau: Environment and Climate

1. The Last Glacial Maximum refers to a time during the last Ice Age, approximately 30,000–19,000 years ago, when glaciers were at the thickest and sea levels were at their lowest.

Chapter 3. Archaeology at Nevasa, Daimabad, and Inamgaon

1. There is one uncalibrated radiocarbon date from the uppermost Chalcolithic layer at Nevasa, obtained from a piece of charcoal. The date was 3106 +/- 122 B.P. (984 B.C.–1228

B.C.). The Chalcolithic deposit ended sometime around 1200–1000 B.C., prior to the Late Jorwe phase at Inamgaon. The rest of these date ranges are derived from chronologies based in culture history—estimates based on ceramic styles and stone tool technology.

2. *Sus* sp. were also infrequent at Inamgaon, but they were common at other sites, such as nearby Chalcolithic Nasik.

3. There are actually 75 individuals, but one individual (VM 75) is from the Indo-Roman layer at the site. This adult individual bears no relevance to the current research and is not considered here but has been fully described previously (Kennedy and Malhotra 1966; Mushrif 2001, 2006).

4. There were originally some 126 individuals, however, approximately 50 individuals are no longer available for analysis (S. R. Walimbe, pers. comm.).

5. Dr. Tripathy and I worked together to sort a few comingled individuals, discuss age estimation for a few problematic individuals, and share radiographs. The observations and interpretations provided in this book were made by me, independent of Dr. Tripathy, and I take full responsibility for the contents, results, conclusions, and any errors herein.

6. The type site of Jorwe also yielded a few sherds of burial urns but no skeletal material.

7. Burials uncovered at these other sites were misplaced and detailed study has not been undertaken. We do know that the Tekwada burials were located in a cemetery on the opposite bank of the river, distinct from the main habitation area (Sali 1986). This settlement plan has parallels at Indus sites Harappa, Kalibangan, and Rakhigarhi.

8. The original numbering system for Nevasa's skeletal collection was deemed to be very confusing and was revised by Veena Mushrif (2001). The details of the original system are provided in her monograph (Mushrif 2006).

9. These three species are presently used in Maharashtra for cart axles, tool handles, agricultural implements, and charcoal. The fruit and bark of *C. fistula* and the leaves of *D. latifolia* also have medicinal value (Vishnu-Mittre et al. 1986).

10. *Z. mauritania* produces a hard wood useful for cart construction and agricultural implements, and its leaves are used today for fodder (Vishnu-Mittre 1981).

11. *P. marsupium* is used today for structural timber, its flowers are useful for fever, its leaves are used as animal fodder, and the gum has medicinal value (Vishnu-Mittre 1981).

12. Despite its common name, the charcoal tree, or *T. orientalis*, does not provide particularly good charcoal. It is useful for rope, twine, cloth, and fishing line. The fruit is edible and the leaves are used for fodder (Vishnu-Mittre 1981).

13. The Pila snail can carry the parasite *Angiostrongylus cantonensis*, a nematode, also known as the rat lungworm. Rats are its main host, but humans can be infected when raw snails are consumed and infection can cause eosinophilic meningonencephalitis, a neurological disorder, and death. Pila can also transmit *Echinostoma sufrartyfex* to humans, as can another aquatic gastropod found at Inamgaon, *Digoniostoma pulchella* (Harinasuta et al. 1987). When consumed, *D. pulchella* can also transmit the trematode *Paryphostomum sufrartyfex* to humans, pigs, and dogs.

Chapter 4. Demography

1. Age for subadults was estimated using long-bone length and dental development (details in Saunders et al. 1995).

Chapter 6. Reconstructing Health at Nevasa, Daimabad, and Inamgaon

1. Estimated for subadults using formulas for the femur distal end (Ruff 2007). These formulas are discussed in some detail in chapter 5.

Bibliography

Allchin, B. 1994. *Living Traditions: Studies in the Ethnoarchaeology of South Asia*. South Asia Publications: New Delhi.

Allchin, B., and F. R. Allchin. 1982. *The Rise of Civilization in India and Pakistan*. Cambridge University Press: Cambridge.

Angel, J. L. 1969. The Bases of Paleodemography. *American Journal of Physical Anthropology* 30:427–37.

Ansari, Z. D. 1988. Lithics. In *Excavations at Inamgaon*, ed. M. K. Dhavalikar, H. D. Sankalia, and H. D. Ansari. Deccan College Post-Graduate Research Institute: Pune. 509–41.

Armelagos, G. J., A. Goodman, and K. H. Jacobs. 1991. The Origins of Agriculture: Population Growth During a Period of Declining Health. *Population and Environment* 13:9–22.

Armelagos, G. J., J. H. Mielke, K. H. Owen, D. P. Van Gerven, J. R. Dewey, and P. E. Mahler. 1972. Bone Growth and Development in Prehistoric Populations from Sudanese Nubia. *Journal of Human Evolution* 1:89–119.

Ashmore, R. H. 1981. Bone Growth and Remodeling as a Measure of Nutritional Stress. *Research Reports, Department of Anthropology, University of Massachusetts* 20 (1981):84–95.

Auerbach, B. M., and C. B. Ruff. 2004. Human Body Mass Estimation: A Comparison of Morphometric and Mechanical Methods. *American Journal of Physical Anthropology* 125:331–42.

Baillie, M. G. L. 1991. Marking in Marker Dates: Towards an Archaeology with Historical Precision. *World Archaeology* 23:233–43.

Banerjee, N. R. 1986. *Report on the Excavations at Nagda, Post-Harappan Chalcolithic Site in District Ujjain, Madhya Pradesh, During 1955–57*. Archaeological Survey of India, Government of India: New Delhi.

Bazarsad, N. 2007. Iron-Deficiency Anemia in Early Mongolian Nomads. In *Ancient Health: Skeletal Indicators of Agricultural and Economic Intensification*, ed. M. Cohen and G. Crane-Kramer. University Press of Florida: Orlando. 250–54.

Biewener, A. A., and J. E. Bertram. 1994. Structural Response of Growing Bone to Exercise and Disuse. *Journal of Applied Physiology* 76:946–55.

Bocquet-Appel, J. P. 2007. *Recent Advances in Paleodemography: Data, Techniques and Patterns*. Springer Verlag: Dordrecht.

Bocquet-Appel, J. P., and C. Masset. 1982. Farewell to Paleodemography. *Journal of Human Evolution* 11:321–33.

———. 1996. Paleodemography: Expectancy and False Hope. *American Journal of Physical Anthropology* 99:571–83.

Bogin, B. 1999. Evolutionary Perspective on Human Growth. *Annual Review of Anthropology* 28:109–53.

Bogin, B., and J. Loucky. 1997. Plasticity, Political Economy, and Physical Growth Status of Guatemalan Children Living in the United States. *American Journal of Physical Anthropology* 102:17–32.

Bogin, B., and R. B. Macvean. 1983. The Relationship of Socioeconomic Status and Sex to Body Size, Skeletal Maturation, and Cognitive Status of Guatemalan City School Children. *Child Development* 54:115–28.

Bourdieu, P. 1984. *Distinction: A Social Critique of the Judgement of Taste*. Harvard University Press: Cambridge.

Bradley, R. S. 1999. *Paleoclimatology: Reconstructing Climates of the Quaternary*. University of Massachusetts Press: Amherst.

Bridges, P. S. 1989. Changes in Activities with the Shift to Agriculture in the Southeastern United States. *Current Anthropology* 30:385–94.

———. 1991a. Degenerative Joint Disease in Hunter-Gatherers and Agriculturalists from the Southeastern United States. *American Journal of Physical Anthropology* 85:379–91.

———. 1991b. Skeletal Evidence of Changes in Subsistence Activities between the Archaic and Mississippian Time Periods in Northwestern Alabama. In *What Mean These Bones: Studies in Southeastern Bioarchaeology*, ed. M. L. Powell, P. S. Bridges, and A. M. W. Mires. University of Alabama Press: Tuscaloosa. 89–101.

———. 1995. Biomechanical Changes in Long Bone Diaphyses with the Intensification of Agriculture in the Lower Illinois Valley. *American Journal of Physical Anthropology Supplement* 20:68.

Bridges, P. S., J. H. Blitz, and M. C. Solano. 2000. Changes in Long Bone Diaphyseal Strength with Horticultural Intensification in West-Central Illinois. *American Journal of Physical Anthropology* 112:217–38.

Brothwell, D. R. 1981. *Digging Up Bones: The Excavation, Treatment and Study of Human Skeletal Remains*. Cornell University Press: Ithaca.

Bryson R. A., and A. M. Swain. 1981. Holocene Variations of Monsoon Rainfall in Rajasthan. *Quaternary Research* 16:135–45.

Buhler G. 1859. On the Hindu God Parjanya. *Transactions of the Philological Society*, December 1860, 154–68.

Buikstra, J. E. 1976. *Hopewell in the Lower Illinois Valley: A Regional Study of Human Biological Variability and Prehistoric Mortuary Behavior*. Northwestern University: Evanston, Ill.

Buikstra, J. E., L. W. Konigsberg, and J. Bullington. 1986. Fertility and the Development of Agriculture in the Prehistoric Midwest. *American Antiquity* 51:528–46.

Butler, N., and E. Alberman. 1969. *Perinatal Problems: The Second Report of the 1958 British Perinatal Survey.* E&S Livingstone: Edinburgh.

Cameron, N. 1979. The Growth of London Schoolchildren, 1904–1966: An Analysis of Secular Trend and Intracounty Variation. *Annals of Human Biology* 6:505–25.

———. 1991. Human Growth, Nutrition, and Health Status in Sub-Saharan Africa. *American Journal of Physical Anthropology* 34 S13:211–50.

———. 2004. Measuring Growth. In *Methods in Human Growth Research*, ed. R. C. Hauspie, N. Cameron, and L. Molinari. Cambridge University Press: Cambridge.

Caratini, C., I. Bentaleb, M. Fontugne, M. T. Morzadec-Kerfourn, J. P. Pascal, and C. Tissot. 1994. A Less Humid Climate since Ca. 3500 Yr B.P. From Marine Cores Off Karwar, Western India. *Palaeogeography, Palaeoclimatology, Palaeoecology* 109:371–84.

Caratini, C., M. Fontugne, J. P. Pascal, C. Tissot, and I. Bentaleb. 1991. A Major Change at Ca. 3500 Years B.P. in the Vegetation of the Western Ghats in North Kanara, Karnataka. *Current Science* 61:669–72.

Carter, D. R., and G. S. Beaupré. 2001. *Skeletal Function and Form: Mechanobiology of Skeletal Development, Aging, and Regeneration.* Cambridge University Press: London.

Carter, D. R., M. C. H. Van der Meulen, and G. S. Beaupré. 1996. Mechanical Factors in Bone Growth and Development. *Bone* 18:S5–S10.

Clemens, S. C., D. W. Murray, and W. L. Prell. 1996. Nonstationary Phase of the Plio-Pleistocene Asian Monsoon. *Science* 274:943–48.

Clutton-Brock, J. 1989. *The Walking Larder: Patterns of Domestication, Pastoralism, and Predation.* Unwin Hyman: London.

Coale, A. J., and P. Demeny. 1983. *Regional Model Life Tables and Stable Populations.* Academic Press: New York.

Cohen, M. N. 1977. *The Food Crisis in Prehistory.* Yale University Press: New Haven.

———. 1989. *Health and the Rise of Civilization.* Yale University Press: New Haven.

Cohen, M. N., and G. J. Armelagos. 1984. *Paleopathology at the Origins of Agriculture.* Academic Press: New York.

Cohen, M. N., and G. G. M. Crane-Kramer. 2007. *Ancient Health: Skeletal Indicators of Agricultural and Economic Intensification.* University Press of Florida: Gainesville.

Cohen, M. N., R. S. Malpass, and H. C. Klein. 1980. *Biosocial Mechanisms of Population Regulation.* Yale University Press: New Haven.

Cole, T. M. 1994. Size and Shape of the Femur and Tibia in Northern Great Plains Indians. In *Skeletal Biology in the Great Plains: Migration, Warfare, Health, and Subsistence*, ed. D. W. Owsley and R. L. Jantz. Smithsonian Institution Press: Washington, D.C. 219–333.

Cook, D. C. 1979. Subsistence Base and Health in Prehistoric Illinois Valley: Evidence from the Human Skeleton. *Medical Anthropology* 3:109–24.

———. 1984. Subsistence and Health in the Lower Illinois Valley: Osteological Evidence. In *Paleopathology at the Origins of Agriculture*, ed. M. N. Cohen and G. Armelagos. Academic Press: New York. 235–69.

Courtillot, V., and J. R. McClinton. 2002. *Evolutionary Catastrophes: The Science of Mass Extinction.* Cambridge University Press: Cambridge.

Cowgill, L. W. 2008. The Ontogeny of Late Pleistocene Postcranial Robusticity. Ph.D. dissertation, Washington University, St. Louis.

———. 2010. The Ontogeny of Holocene and Late Pleistocene Human Postcranial Strength. *American Journal of Physical Anthropology* 141:16–37.

Cowgill, L. W., and L. Hager. 2007. Variation in the Development of Postcranial Robusticity: An Example from Çatalhöyük, Turkey. *International Journal of Osteoarchaelogy* 17:235–52.

Cowin, S. C. 2001. *Bone Mechanics Handbook*. CRC Press: Boca Raton, Fla.

Currey, J. D. 2002. *Bones: Structure and Mechanics*. Princeton University Press: Princeton.

Daniels, R. J. 1997. Taxonomic Uncertainties and Conservation Assessment of the Western Ghats. *Current Science* 73:169–70.

Das, P. K. 1995. *The Monsoons*. National Book Trust of India: New Delhi.

Dasog, G. S. 2002. Black Soils of India: A Survey of Pedological Research since 1947. In *Archaeology and Interactive Disciplines*, ed. S. Settar and R. Korisettar. Manohar: Delhi. 1–22.

Demes, B. 2007. In Vivo Bone Strain and Bone Functional Adaptation. *American Journal of Physical Anthropology* 133:717–22.

Dhavalikar, M. K. 1977. Inamgaon: The Pattern of Settlement. *Man and Environment* 1:46–51.

———. 1984. Toward an Ecological Model for Chalcolithic Cultures of Central and Western India. *Journal of Anthropological Archaeology* 3:133–58.

———. 1985. Cultural Ecology of Chalcolithic Maharashtra. In *Recent Advances in Indian Archaeology*, ed. S. B. Deo and K. Paddaya. Deccan College: Poona. 65–73.

———. 1988. *The First Farmers of the Deccan*. Ravish: Pune.

———. 1989. Human Ecology in Western India in the Second Millennium B.C. *Man and Environment* 14:83–90.

———. 1994a. Chalcolithic Architecture at Inamgaon and Walki: An Ethnoarchaeological Study. In *Living Traditions: Studies in the Ethnoarchaeology of South Asia*, ed. B. Allchin. Oxford and IBH Publishing: New Delhi. 31–52.

———. 1994b. Early Farming Communities of Central India. *Man and Environment* 19:159–68.

———. 1997. *Indian Protohistory*. Books and Books: New Delhi.

———. 2004. The Nile Floods and Indian Monsoon. In *Monsoon and Civilization*, ed. Y. Yasuda, V. S. Shinde, and S. Kenkyu. Roli Books: New Delhi. 207–12.

Dhavalikar, M. K., and Z. D. Ansari. 1988. Other Artifacts. In *Excavations at Inamgaon*, ed. M. K. Dhavalikar. Deccan College Post-Graduate Research Institute: Pune. 554–726.

Dhavalikar, M. K., and G. L. Possehl. 1974. Subsistence Pattern of an Early Farming Community of Western India. *Puratattva*:39–46.

Dhavalikar, M. K., H. D. Sankalia, and Z. D. Ansari. 1988. *Excavations at Inamgaon*. Deccan College Post-Graduate Research Institute: Pune.

Diamond, J. 2004. *Collapse: How Societies Choose to Fail or Succeed*. Viking Books: New York.

Domett, K. M., and N. Tayles. 2006. Adult Fracture Patterns in Prehistoric Thailand: A Biocultural Interpretation. *International Journal of Osteoarchaeology* 16:185–99.

———. 2007. Population Health from the Bronze to the Iron Age in the Mun River Valley, Northeast Thailand. In *Ancient Health: Skeletal Indicators of Agricultural and Economic Intensification*, ed. M. Cohen and G. Crane-Kramer. University Press of Florida: Orlando.

Douglas, M. T. 2006. Subsistence Change and Dental Health in the Past People of Non Nok Tha, Northeast Thailand. In *Bioarchaeology of Southeast Asia*, ed. M. Oxenham and N. Tayles. Cambridge University Press: Cambridge. 191–219.

Douglas, M. T., and M. Pietrusewsky. 2007. Biological Consequences of Sedentism and Agricultural Intensification in Northeast Thailand. In *Ancient Health: Skeletal Indicators of Agricultural and Economic Intensification*, ed. M. N. Cohen and G. Crane-Kramer. University Press of Florida: Orlando. 300–319.

Duncan, G. 1880. *Geography of India: Comprising a Descriptive Outline of All India and a Detailed Geographical, Commercial, Social, and Political Account of Each of Its Provinces.* Higgenbotham: Madras.

Ellison, P. 2001. *On Fertile Ground: A Natural History of Human Reproduction.* Harvard University Press: Cambridge.

Enzel, Y., L. L. Ely, S. Mishra, R. Ramesh, R. Amit, B. Lazar, S. N. Rajaguru, V. R. Baker, and A. Sandler-Ericksen. 1990. High-Resolution Holocene Environmental Changes in the Thar Desert, Northwestern India. *Science* 284:125–28.

Ericksen, M. F. 1976. Cortical Bone Loss with Age in Three Native American Populations. *American Journal of Physical Anthropology* 45:443–52.

Evelyth, P. B., and J. M. Tanner. 1990. *Worldwide Variation in Human Growth.* Cambridge University Press: Cambridge.

Fairservis, W. A. 1979. The Origin, Character and Decline of an Early Civilization. *Ancient Cities of the Indus*:66–89.

Fazekas, I. G., and F. Kósa. 1978. *Forensic Fetal Osteology.* Akadémiai Kiadó: Budapest.

Feldesman, M. R. 1992. Femur Stature Ratio and Estimates of Stature in Children. *American Journal of Physical Anthropology* 87:447–59.

Floyd, B. 2002. Evidence of Age-Related Responses to Short-Term Environmental Variation: Time Series Analysis of Cross-Sectional Data from Taiwan, 1969 to 1990. *American Journal of Human Biology* 14:61–73.

Fogel, R. W. 1986. Physical Growth as a Measure of the Economic Well Being of Populations: The Eighteenth and Nineteenth Centuries. In *Human Growth*, ed. F. Faulkner and J. M. Tanner. Plenum: New York. 263–81.

Froment, A. 2001. Evolutionary Biology and Health of Hunter-Gatherer Populations. In *Hunter-Gatherers: An Interdisciplinary Perspective*, ed. C. Panter-Brick, R. H. Layton, and P. Rowley-Conway. Cambridge University Press: Cambridge. 239–66.

Fuller, D., R. Korisettar, P. C. Venkatasubbaiah, and M. K. Jones. 2004. Early Plant Domestications in Southern India: Some Preliminary Archaeobotanical Results. *Vegetation History and Archaeobotany* 13:115–29.

Fuller, D. Q. 2003. An Agricultural Perspective on Dravidian Historical Linguistics: Archaeological Crop Packages, Livestock and Dravidian Crop Vocabulary. In *Exam-

ining the Farming/Language Dispersal Hypothesis, ed. P. Bellwood and C. Renfrew McDonald. Institute for Archaeological Research: Cambridge.

———. 2006. Agricultural Origins and Frontiers in South Asia: A Working Synthesis. *Journal of World Prehistory* 20:1–86.

———. 2007. Contrasting Patterns in Crop Domestication and Domestication Rates: Recent Archaeobotanical Insights from the Old World. *Annals of Botany* 100:1–22.

Gage, T. B. 2000. Demography. In *Human Biology: An Evolutionary and Biocultural Perspective*, ed. S. Stinson, B. Bogin, R. Huss-Ashmore, and D. O'Rourke. Wiley: New York. 507–51.

Garn, S. M. 1970. *The Earlier Gain and the Later Loss of Compact Bone*. Thomas: Springfield, Ill.

———. 1980. Human Growth. *Annual Review of Anthropology* 9:175–292.

Genovese, S. 1967. Proportionality of the Long Bones and Their Relation to Stature among Mesoamericans. *American Journal of Physical Anthropology* 26:67–78.

Gogte, V. D., and A. Kshirsagar. 1988. Archaeochemistry. In *Excavations at Inamgaon*, ed. M. K. Dhavalikar, H. D. Sankalia, and Z. D. Ansari. Deccan College Post-Graduate Research Institute: Pune.

Goodman, A. H., G. J. Armelagos, and J. C. Rose. 1980. Enamel Hypoplasias as Indicators of Stress in Three Prehistoric Populations from Illinois. *Human Biology* 52:515–28.

Goodman, A. H., J. Lallo, G. J. Armelagos, and J. C. Rose. 1984. Health Changes at Dickson Mounds, Illinois (A.D. 950–1300). In *Paleopathology at the Origins of Agriculture*, ed. M. N. Cohen and G. Armelagos. University Press of Florida: Gainesville. 271–305.

Goodman, A. H., and T. L. Leatherman. 1999. *Building a New Biocultural Synthesis: Political-Economic Perspectives on Human Biology*. University of Michigan Press: Ann Arbor.

Goodman, A. H., and D. L. Martin. 2002. Health Profiles from Skeletal Remains. In *The Backbone of History: Health and Nutrition in the Western Hemisphere*, ed. R. H. Steckel and J. Rose. Cambridge University Press: Cambridge.

Goodman, A. H., R. B. Thomas, A. C. Swedlund, and G. J. Armelagos. 1988. Biocultural Perspectives on Stress in Prehistoric, Historical, and Contemporary Population-Research. *Yearbook of Physical Anthropology* 31:169–202.

Gowland, R. L., and A. T. Chamberlain. 2002. A Bayesian Approach to Ageing Perinatal Skeletal Material from Archaeological Sites: Implications for the Evidence for Infanticide in Roman-Britain. *Journal of Archaeological Science* 29:677–85.

Gunnell, D., J. Rogers, and P. Dieppe. 2001. Height and Health: Predicting Longevity from Bone Length in Archaeological Remains. *Journal of Epidemiology and Community Health* 55:505–7.

Halcrow, S., and N. Tayles. 2008. Stress Near the Start of Life? Localized Hypoplasia of the Primary Canine in Late Prehistoric Mainland Southeast Asia. *Journal of Archaeological Science* 35:2215–22.

Halcrow, S., N. Tayles, and V. Livingstone. 2008. Infant Death in Late Prehistoric Southeast Asia. *Asian Perspectives* 47:371–404.

Harinasuta, T., D. Bunnag, and P. Radomyos. 1987. Intestinal Fluke Infections. *Bailliere's Clinical Tropical Medicine and Communicable Diseases* 2:695–721.

Harris, D. R., and V. G. Childe. 1992. *The Archaeology of V. Gordon Childe: Contemporary Perspectives*. University College, London Institute of Archaeology: London. Reprint, University of Chicago Press: Chicago.

Haviland, W. A. 1967. Stature at Tikal, Guatemala: Implications for Ancient Maya Demography and Social Organization. *American Antiquity* 32:316–25.

Himes, J. H., R. Martorell, J. P. Habicht, C. Yarbrough, R. M. Malina, and R. E. Klein. 1975. Patterns of Cortical Bone Growth in Moderately Malnourished Preschool Children. *Human Biology* 47:337–50.

Himes, J. H., C. Yarbrough, and R. Martorell. 1977. Estimation of Stature in Children from Radiographically Determined Metacarpal Length. *Journal of Forensic Sciences* 22.

Hinton, R. J., M. O. Smith, and F. H. Smith. 1980. Tooth Size Changes in Prehistoric Tennessee Indians. *Human Biology* 52:229–45.

Hobbes, T., and C. B. Macpherson. 1968. *Leviathan*. Penguin Books: Baltimore.

Holliday, T. W. 1995. *Body Size and Proportions in the Late Pleistocene Western Old World and the Origins of Modern Humans*. Albuquerque: University of New Mexico.

———. 1997. Postcranial Evidence of Cold Adaptation in European Neandertals. *American Journal of Physical Anthropology* 104:245–58.

Hooja, R. 1988. *The Ahar Culture and Beyond: Settlements and Frontiers of "Mesolithic" and Early Agricultural Sites in South-Eastern Rajasthan, C. 3rd–2nd Millennia B.C.* BAR: Oxford.

Hoppa, R. D. 1992. Evaluating Human Skeletal Growth: An Anglo-Saxon Example. *International Journal of Osteoarchaeology* 2:275–88.

Hoppa, R. D., and C. M. Fitzgerald. 1999. *Human Growth in the Past: Studies from Bones and Teeth*. Cambridge University Press: Cambridge.

Hoppa, R. D., and J. W. Vaupel. 2002. *Paleodemography: Age Distributions from Skeletal Samples*. Cambridge University Press: New York.

Horowitz, S., G. Armelagos, and K. Wachter. 1988. On Generating Birth Rates from Skeletal Populations. *American Journal of Physical Anthropology* 76:189–96.

Hrdy, S. B. 1999. *Mother Nature: Maternal Instincts and How They Shape the Human Species*. Ballantine Books: New York.

Hummert, J. R. 1983. Cortical Bone Growth and Dietary Stress among Subadults from Nubia's Batn El Hajar. *American Journal of Physical Anthropology* 62:167–76.

Hummert, J. R., and D. P. Van Gerven. 1983. Skeletal Growth in a Medieval Population from Sudanese Nubia. *American Journal of Physical Anthropology* 60:471–78.

Huss-Ashmore, R. 2000. Theory in Human Biology: Evolution, Ecology, Adaptability, and Variation. In *Human Biology: An Evolutionary and Biocultural Perspective*, ed. S. Stinson, B. Bogin, R. Huss-Ashmore, and D. O'Rourke. Wiley: New York.

Huss-Ashmore, R., and F. E. Johnston. 1985. Bioanthropological Research in Developing Countries. *Annual Review of Anthropology* 14:475–28.

Hutchinson, D. L. 2002. *Foraging, Farming, and Coastal Biocultural Adaptation in Late Prehistoric North Carolina*. University Press of Florida: Gainesville.

———. 2004. *Bioarchaeology of the Florida Gulf Coast: Adaptation, Conflict, and Change*. University Press of Florida: Gainesville.

India, Government of. 2004. Agricultural Statistics at a Glance. In *Directorate of Economics and Statics*, ed. Indian Farmers Fertilizer Cooperative Unlimited.

Indian Archaeology—A Review [IAR]. 1956. In *Indian Archaeology 1955–1956*, ed. A. Ghosh. Department of Archaeology, Government of India: Delhi. 1–20.

———. 1961. In *Indian Archaeology 1960–61*, ed. A. Ghosh. Department of Archaeology, Government of India: Delhi. 17–18.

———. 1962. In *Indian Archaeology 1961–1962*, ed. A. Ghosh. Department of Archaeology, Government of India: Delhi. 24–25.

———. 1963. In *Indian Archaeology 1961–62*, ed. A. Ghosh. Department of Archaeology, Government of India: Delhi. 11–12.

———. 1964. In *Indian Archaeology 1963–64*, ed. A. Ghosh. Department of Archaeology, Government of India: Delhi. 15–16.

———. 1965. In *Indian Archaeology 1964–65*, ed. A. Ghosh. Department of Archaeology, Government of India: Delhi. 16–18.

———. 1988. In *Indian Archaeology 1987–88*, ed. M. C. Joshi. Department of Archaeology, Government of India: Delhi. 77–78.

Jackes, M. K. 1986. The Mortality of Ontario Archaeological Populations. *Canadian Journal of Anthropology* 5:33–48.

———. 1992. Paleodemography: Problems and Methods. In *Skeletal Biology of Past Peoples: Research Methods*, ed. S. Saunders and A. Katzenberg. Wiley-Liss: New York. 189–224.

Jackes, M., and C. Meiklejohn. 2004. Building a Method for the Study of the Mesolithic-Neolithic Transition in Portugal. *Documenta Praehistorica* 31:89–111.

Jain, M. 2000. Stratigraphic Development of Some Exposed Quaternary Alluvial Sequences in the Thar and Its Margins: Fluvial Response to Climate Change, Western India. Ph.D. dissertation, University of Delhi.

Jain, M., and S. K. Tandon. 2003. Fluvial Response to Late Quaternary Climate Changes, Western India. *Quaternary Science Reviews* 22:2223–35.

Jantz, R. L., and D. W. Owsley. 1984. Long-Bone Growth Variation among Arikara Skeletal Populations. *American Journal of Physical Anthropology* 63:13–20.

Johansson, S. R., and D. Owsley. 2002. Welfare History on the Great Plains: Mortality and Skeletal Health, 1650–1900. In *The Backbone of History: Health and Nutrition in the Western Hemisphere*, ed. R. H. Steckel and J. Rose. Cambridge University Press: New York.

Johnston, F. E. 1962. Growth of Long Bones of Infants and Young Children at Indian Knoll. *American Journal of Physical Anthropology* 20:249–54.

Joshi, V. U., and V. Kale. 1997. Colluvial Deposits in Northwest Deccan, India: Their Significance in the Interpretation of Late Quaternary History. *Journal of Quaternary Science* 12:391–403.

Jungers, W. 1988. Lucy's Length: Stature Reconstruction in Australopithecus Afarensis (A.L.288-1) with Implications for Other Small-Bodied Hominids. *American Journal of Physical Anthropology* 76:227–31.

Kajale, M. D. 1988. Plant Remains. In *Excavations at Inamgaon*, ed. M. K. Dhavalikar,

H. D. Sankalia, and Z. D. Ansari. Deccan College Post-Graduate Research Institute: Pune. 727–32.

———. 1990. Observations on the Plant Remains from Excavation at Chalcolithic Kaothe, District Dhule, Maharashtra with Cautionary Remarks on Their Interpretations. In *Excavations at Kaothe*, ed. M. K. Dhavalikar et al. Deccan College Post-Graduate Research Institute: Pune. 265–80.

Kajale, M. D., G. L. Badam, and S. N. Rajaguru. 1976. Late Quaternary History of the Ghod Valley, Maharashtra. *Geophytology* 6:122–32.

Kale, V. S. 1999. Late Holocene Temporal Patterns of Paleofloods in Central and Western India. *Man and Environment* 24:109–15.

———. 2002. Fluvial Geomorphology of Indian Rivers: An Overview. *Progress in Physical Geography* 26:400–433.

———. 2007. Fluvio-Sedimentary Response of the Monsoon-Fed Indian Rivers to Late Pleistocene-Holocene Changes in Monsoon Strength: Reconstruction Based on Existing 14c Dates. *Quaternary Science Reviews* 26:1610–20.

Kale, V. S., and S. N. Rajaguru. 1987. Late Quaternary Alluvial History of the Northwestern Deccan Upland Region. *Nature* 325:612–14.

———. 1988. Morphology and Denudation Chronology of the Coastal and Upland River Basins of Western Deccan Trappean Landscape (India): A Collation. *Memoirs—Geological Society of India* 47:333–48.

Katzenberg, M. A., and S. R. Saunders. 2000. *Biological Anthropology of the Human Skeleton*. Wiley-Liss: New York.

Keith, M. S. 1981. Cortical Bone Loss in Juveniles of Dickson Mounds. *Research Reports*:64–77.

Kennedy, K. A. R. 1984. Growth, Nutrition, and Pathology in Changing Paleodemographic Settings in South Asia. In *Paleopathology at the Origins of Agriculture*, ed. M. N. Cohen and G. Crane-Kramer. University Press of Florida: Orlando. 169–92.

Kennedy, K. A. R., N. C. Lovell, and C. B. Burrow. 1986. *Mesolithic Human Remains from the Gangetic Plain: Sarai Nahar Rai*. Cornell University: Ithaca.

Kennedy, K. A. R., J. R. Lukacs, R. F. Pastor, T. L. Johnston, N. C. Lovell, J. N. Pal, B. E. Hemphill, and C. B. Burrow. 1992. *Human Skeletal Remains from Mahadaha: A Gangetic Mesolithic Site*. Cornell University Press: Ithaca.

Kennedy, K. A. R., and K. C. Malhotra. 1966. Human Skeletal Remains from Chalcolithic and Indo-Roman Levels from Nevasa: An Anthropometric and Comparative Analysis. In *Deccan College Building Centenary and Silver Jubilee Series*. Deccan College: Poona.

Kennett, D. K., and B. Winterhalder. 2006. *Behavioral Ecology and the Transition to Agriculture*. University of California Press: Berkeley and Los Angeles.

Kieser, J. A. 1990. *Human Adult Odontometrics: The Study of Variation in Adult Tooth Size*. Cambridge University Press: Cambridge.

Konigsberg, L. W., and S. R. Frankenberg. 1992. Estimation of Age Structure in Anthropological Demography. *American Journal of Physical Anthropology* 89:235–56.

Krigbaum, J. 2007. Prehistoric Dietary Transitions in Tropical Southeast Asia: Stable Isotope and Dental Caries Evidence from Two Sites in Malaysia. In *Ancient Health:*

Skeletal Indicators of Agricultural and Economic Intensification, ed. M. N. Cohen and G. Crane-Kramer. University Press of Florida: Orlando. 273–85.

Kumaran, K. P. N., K. M. Nair, M. Shindikar, R. B. Limaye, and D. Padmalal. 2005. Stratigraphical and Palynological Appraisal of the Late Quaternary Mangrove Deposits of the West Coast of India. *Quaternary Research* 64:418–31.

Lallo, J. 1973. The Skeletal Biology of Three Prehistoric Amerindian Populations from Dickson Mounds. Ph.D. dissertation, University of Massachusetts, Amherst.

Lambert, P. M. 2000. *Bioarchaeological Studies of Life in the Age of Agriculture: A View from the Southeast*. University of Alabama Press: Tuscaloosa.

Lanyon, L., and T. Skerry. 2001. Postmenopausal Osteoporosis as a Failure of Bone's Adaptation to Functional Loading: A Hypothesis. *Journal of Bone Mineral Research* 16:1937–47.

Larsen, C. S. 1982. *The Anthropology of St. Catherines Island: Part 3. Prehistoric Human Biological Adaptation*. American Museum of Natural History: New York.

———. 1995. Biological Changes in Human Populations with Agriculture. *Annual Review of Anthropology* 24:185–213.

Larsen, C. S., C. B. Ruff, and M. C. Griffin. 1996. Implications of Changing Biomechanical and Nutritional Environments for Activity and Lifeway in the Eastern Spanish Borderlands. In *Bioarchaeology of Native American Adaptation in the Spanish Borderlands*, ed. B. J. Baker et al. University Press of Florida: Orlando. 95–125.

Last, J. M. 2001. *A Dictionary of Epidemiology*. Oxford University Press: Oxford.

Lieberman, D. E., M. J. Devlin, and O. M. Pearson. 2001. Articular Area Responses to Mechanical Loading: Effects of Exercise, Age, and Skeletal Location. *American Journal of Physical Anthropology* 116:266–77.

Lieberman, D. E., J. D. Polk, and B. Demes. 2004. Predicting Long Bone Loading from Cross-Sectional Geometry. *American Journal of Physical Anthropology* 123:156–71.

Livi-Bacci, M. 2007. *A Concise History of World Population*. Blackwell: Oxford.

Lovejoy, C. O., K. F. Russell, and M. L. Harrison. 1990. Long Bone Growth Velocity in the Libben Population. *American Journal of Human Biology* 2:533–41.

Lucy, D., R. G. Aykroyd, A. M. Pollard, and T. Solheim. 1996. A Bayesian Approach to Adult Human Age Estimation from Dental Observations by Johanson's Age Changes. *Journal of Forensic Science* 41:189–94.

Lukacs, J. R. 1980. Paleodemography in Prehistoric India—Mortality and Morbidity at Post-Harappan Inamgaon. *American Journal of Physical Anthropology* 52:250.

———. 1983. Dental Anthropology and the Origins of Two Iron Age Populations from Northern Pakistan. *Homo*:1–15, ill.

———. 1992. Dental Paleopathology and Agricultural Intensification in South Asia: New Evidence from Bronze Age Harappa. *American Journal of Physical Anthropology* 87:133–50.

———. 1997. New Frontiers in Dental Anthropology: Creative Approaches to Diet and Stress in Prehistory. *Biological Anthropology: The State of the Science*:117–30.

———. 1999. Interproximal Contact Hypoplasia in Primary Teeth: A New Enamel Defect with Anthropological and Clinical Relevance. *American Journal of Human Biology* 11:718–34.

———. 2002. Hunting and Gathering Strategies in Prehistoric India: A Biocultural Perspective on Trade and Subsistence. In *Forager Traders in South and Southeast Asia: Long Term Histories*, ed. K. Morrison and L. Junker. Cambridge University Press: Cambridge. 41–61.

———. 2008. Fertility and Agriculture Accentuate Sex Differences in Dental Caries Rates. *Current Anthropology* 49:901–14.

———. 2009. Markers of Physiological Stress in Juvenile Bonobos (Pan Paniscus): Are Enamel Hypoplasia, Skeletal Development and Tooth Size Interrelated? *American Journal of Physical Anthropology* 139:339–52.

Lukacs, J. R., and G. L. Badam. 1981. Paleodemography of Post-Harappan Inamgaon: A Preliminary Report. *Journal of the Indian Anthropological Society* 16:59–74.

Lukacs, J. R., R. K. Bogorad, S. R. Walimbe, and D. C. Dunbar. 1986. Paleopathology at Inamgaon: A Post-Harappan Agrarian Village in Western India. *Proceedings of the American Philosophical Society* 130:289–311.

Lukacs, J. R., and L. L. Minderman. 1992. Dental Pathology and Agricultural Intensification from Neolithic to Chalcolithic Periods at Mehrgarh (Baluchistan, Pakistan). In *South Asian Archaeology 1989: Papers from the Tenth International Conference of South Asian Archaeologists in Western Europe, Musée national des Arts asiatiques— Guimet, Paris, France, 3–7 July 1989*, ed. C. Jerrige, J. P. Gerry, and R. H. Meadow. Prehistory Press: Madison, WI. 167–79.

Lukacs, J. R., G. C. Nelson, and S. R. Walimbe. 2001. Enamel Hypoplasia and Childhood Stress in Prehistory: New Data from India and Southwest Asia. *Journal of Archaeological Science* 28:1159–69.

Lukacs, J. R., and J. N. Pal. 1992. Dental Anthropology of Mesolithic Hunter-Gatherers: A Preliminary Report on the Mahadaha and Sarai Nahar Rai Dentition. *Man and Environment* 17:45–55.

———. 1993. Mesolithic Subsistence in North India: Inferences from Dental Pathology and Odontometry. *Current Anthropology* 34:745–65.

———. 2004. Paleopathology and Subsistence Transition Theory: New Evidence from Mesolithic Skeletons from Damdama. In *Dr R K Varma Felicitation Volume*, ed. J. N. Pal. Swabha Prakashan: Allahabad.

Lukacs, J. R., and S. R. Walimbe. 1984. Paleodemography at Inamgaon: An Early Farming Village in Western India. In *The People of South Asia: The Biological Anthropology of India, Pakistan, and Nepal*, ed. J. R. Lukacs. Plenum Press: New York. 105–32.

———. 1986. *Excavations at Inamgaon. Vol. 2, The Physical Anthropology of Human Skeletal Remains: An Osteobiogeographic Analysis.* Deccan College Post-Graduate and Research Institute: Pune.

———. 1998. Physiological Stress in Prehistoric India: New Data on Localized Hypoplasia of Primary Canines Linked to Climate and Subsistence Change. *Journal of Archaeological Science* 25:571–85.

———. 2000. Health, Climate, and Culture in Prehistoric India: Conflicting Conclusions from Archaeology and Anthropology? In *IsMEO* (Istituto italiano Medio Estremo Oriente). Naples. 363–81.

———. 2005. Biological Responses to Subsistence Transitions in Prehistory: Diachronic Dental Changes at Chalcolithic Inamgaon. *Man & Environment* 30:24–43.

———. 2007a. Climate, Subsistence and Health in Prehistoric India: The Biological Impact of a Short-Term Reversal. In *Ancient Health: Skeletal Indicators of Agricultural and Economic Intensification*, ed. M. N. Cohen and G. Crane-Kramer. University Press of Florida: Gainesville.

———. 2007b. Interpreting Biological Diversity in South Asian Prehistory: Early Holocene Population Affinities and Subsistence Adaptations. In *The Evolution and History of Human Populations in South Asia*, ed. M. D. Petraglia and B. Allchin. Springer Verlag: Dordrecht. 271–96.

Lukacs, J. R., S. R. Walimbe, and B. Floyd. 2001. Epidemiology of Enamel Hypoplasia in Deciduous Teeth: Explaining Variation in Prevalence in Western India. *American Journal of Human Biology* 13:788–807.

Maresh, M. M. 1959. Linear Body Proportions: A Roentgenographic Study. *American Journal of Diseases in Children* 98:27–49.

———. 1970. Measurements from Roentgenograms. In *Human Growth and Development*, ed. R. W. McCannon. Charles C. Thomas: Springfield, Ill. 157–88.

Martin, D. L., and A. H. Goodman. 2002. Health Conditions before Columbus: Paleopathology of Native North Americans. *Western Journal of Medicine* 176:65–68.

Mays, S. 1995. Relationship between Harris Lines and Other Aspects of Skeletal Development in Adults and Juveniles. *Journal of Archaeological Science* 22:511–20.

Mays, S. A. 1999. Osteoporosis in Earlier Human Populations. *Journal of Clinical Densitometry* 2:71–78.

McCaa R. 1998. *Calibrating Paleodemography: The Uniformitarian Challenge Turned.* American Association of Physical Anthropology annual meeting. Salt Lake City.

———. 2002. Paleodemography of the Americas. In *The Backbone of History: Health and Nutrition in the Western Hemisphere*, ed. R. H. Steckel and J. Rose. Cambridge University Press: New York. 94–126.

McEwan, J. M., S. Mays, and G. M. Blake. 2005. The Relationship of Bone Mineral Density and Other Growth Parameters to Stress Indicators in a Medieval Juvenile Population. *International Journal of Osteoarchaeology* 15:155–63.

McHenry, H. M. 1994. Behavioral Ecological Implications of Early Hominid Body Size. *Journal of Human Evolution* 27:77–87.

Meher-Homji, V. M. 1989. History of Vegetation of Peninsular India. *Man and Environment* 13:1–10.

Mehra, K. L. 1999. Subsistence Changes in India and Pakistan: The Neolithic and Chalcolithic from the Point of View of Plant Use to-Day. In *The Prehistory of Food*, ed. C. Gosden and J. G. Hather. Rutledge: London. 139–46.

Mehra, K. L., and R. K. Arora. 1985. Some Considerations on the Domestication of Plants in India. *Recent Advances in Indo Pacific Prehistory*:275–79.

Meindl, R. S., and K. F. Russell. 1998. Recent Advances in Method and Theory in Paleodemography. *Annual Review of Anthropology* 27:375–99.

Mensforth, R. P., and C. O. Lovejoy. 1985. Anatomical, Physiological, and Epidemiological Correlates of the Aging Process: A Confirmation of Multifactorial Age-Determination in the Libben Skeletal Population. *American Journal of Physical Anthropology* 68:87–106.

Merchant, V. A., and D. C. Ubelaker. 1977. Skeletal Growth of the Proto-Historic Arikara. *American Journal of Physical Anthropology* 46:61–72.

Misra, V. N. 1997. Balathal: A Chalcolithic Settlement in Mewar, Rajasthan, India: Results of First Three Seasons Excavations. *South Asian Archaeology* 13:251–73.

———. 2005. Radiocarbon Chronology of Balathal, District Udaipur. *Man and Environment* 30:54–61.

Molleson, T. I. 1989. Social Implication of Mortality Patterns of Juveniles from Poundbury Camp, Romano-British Cemetery. *Anthropologischer Anzeiger* 47:27–38.

Moorrees, C. F. A., E. A. Fanning, and E. Hunt. 1963. Age Variation of Formation Stages for Ten Permanent Teeth. *Journal of Dental Research* 42:1490–1502.

Moro, M., M. C. H. Van der Meulen, B. J. Kiratli, R. Marcus, L. K. Bachrach, and D. R. Carter. 1996. Body Mass Is the Primary Determinant of Midfemoral Bone Acquisition During Adolescent Growth. *Bone* 19:519–26.

Morrison, K. D. 2002. Historicizing Adaptation, Adapting to History: Forager-Traders in South and Southeast Asia. In *Forager Traders in South and Southeast Asia: Long Term Histories*, ed. K. D. Morrison and L. L. Junker. Cambridge: Cambridge University Press.

Morrison, K. D., and L. L. Junker, eds. 2002. *Forager-Traders in South and South-East Asia: Long Term Histories*. Cambridge: Cambridge University Press.

Mushrif, V. 2001. Human Remains from Nevasa: An Osteobiographic Analysis. Ph.D. dissertation, Deccan College Post-Graduate Research Institute, Pune.

Mushrif, V., and S. R. Walimbe. 2006. *Human Skeletal Remains from Chalcolithic Nevasa: Osteobiographic Analysis*. British Archaeological Reports.

Must, A., G. E. Dalal, and W. H. Dietz. 1991. Reference Data for Obesity: 85th and 95th Percentiles of Body Mass Index (Wt/Ht2) and Triceps Fold Thickness. *American Journal of Clinical Nutrition* 53:839–46.

Nelson, G. C., J. R. Lukacs, and P. Yule. 1999. Dates, Caries, and Early Tooth Loss During the Iron Age of Oman. *American Journal of Physical Anthropology* 108:333–43.

Ollier, C. D., and K. B. Powar. 1985. The Western Ghats and the Morphotectonics of Peninsular India. *Zeitschrift für Geomorphologie Supplement Bd.* 54:57–69.

O'Neill, M. C., and C. Ruff. 2004. Estimating Human Long Bone Cross-Sectional Geometric Properties: A Comparison of Noninvasive Methods. *Journal of Human Evolution* 47:221–35.

Ortner, D. J. 1991. Theoretical and Methodological Issues in Paleopathology. In *Human Paleopathology: Current Syntheses and Future Options*, ed. D. J. Ortner and A. C. Auferheide. Smithsonian Institution Press: Washington, D.C. 5–11.

Owsley, D. W. 1991. Temporal Variation in Femoral Cortical Thickness of North American Plains Indians. In *Human Palaeopathology: Current Syntheses and Future Options*, ed. D. J. Ortner and A. C. Auferheide. Smithsonian Institution Press: Washington, D.C. 105–10.

Owsley, D. W., and R. L. Jantz. 1985. Long Bone Lengths and Gestational Age Distributions of Post-Contact Period Arikara Indian Perinatal Infant Skeletons. *American Journal of Physical Anthropology* 68:321–28.

Oxenham, M. F., K. T. Nguyen, and L. C. Nguyen. 2006. The Oral Health Consequences of the Adoption and Intensification of Agriculture in Southeast Asia. In *Bioarchaeology of Southeast Asia*, ed. M. Oxenham and N. Tayles. Cambridge University Press: Cambridge. 263–89.

Oxenham, M. F., and N. Tayles. 2006. *Bioarchaeology of Southeast Asia*. Cambridge University Press: Cambridge.

Oxenham, M. F., N. K. Thuy, and N. L. Cuong. 2005. Skeletal Evidence for the Emergence of Infectious Disease in Bronze and Iron Age Northern Vietnam. *American Journal of Physical Anthropology* 126:359–76.

Paine, R. R. 1997. *Integrating Archaeological Demography: Multidisciplinary Approaches to Prehistoric Population*. SIU Palfi: Carbondale, Ill.

Paine, R. R., and H. C. Harpending. 1996. The Reliability of Paleodemographic Fertility Estimators. *American Journal of Physical Anthropology* 101:151–60.

Pandey, D. N., A. K. Gupta, and D. M. Anderson. 2003. Rainwater Harvesting as an Adaptation to Climate Change. *Current Science* 85:46–59.

Panja, S. 1996. Mobility Strategies, Site Structure and Settlement Organization: An Actualistic Perspective. *Man and Environment* 21:58–73.

———. 1999. Mobility and Subsistence Strategies: A Case Study of Inamgaon, a Chalcolithic Site in Western India. *Asian Perspectives* 38:154–85.

———. 2003. Mobility Strategies and Site Structure: A Case Study of Inamgaon. *Journal of Anthropological Archaeology* 22:105–25.

Pappu, R. S. 1988. Site Catchment Analysis. In *Excavations at Inamgaon*, ed. M. K. Dhavalikar, H. D. Sankalia, and Z. D. Ansari. Deccan College Post-Graduate Research Institute: Pune. 107–20.

Partridge, T. C., G. C. Bone, C. J. H. Hartnady, P. B. Demenocal, and W. F. Ruddiman. 1996. Climatic Effects of Late Neogene Tectonism and Volcanism. In *Paleoclimate and Evolution with Emphasis on Human Origins*, ed. E. Vrba, G. H. Denton, T. C. Partridge, and L. H. Burckle. Yale University Press: New Haven.

Pawankar, S. J. 1996. Archaeozoology at Inamgaon. Ph.D. dissertation, University of Poona, Pune.

———. 1997. Man and Animal Relationship in Early Farming Communities of Western India, with Special Reference to Inamgaon. Ph.D. dissertation, Deccan College Post-Graduate Research Institute, Pune.

Pawankar, S. J., and P. K. Thomas. 1997. Fauna and Subsistence Pattern in the Chalcolithic Culture of Western India, with Special Reference to Inamgaon. *Anthropozoologica* 25–26:737–46.

Pearson, O. M. 2000. Activity, Climate, and Postcranial Robusticity: Implications for Modern Human Origins and Scenarios of Adaptive Change. *Current Anthropology* 41:569–607.

Pearson, O. M., and D. E. Lieberman. 2004. The Aging of Wolff's "Law": Ontogeny and Responses to Mechanical Loading Cortical Bone. *Yearbook of Physical Anthropology* 47:63–99.

Pechenkina, E. A., R. A. Benfer, and X. Ma. 2007. Diet and Health in the Neolithic of the Wei and Middle Yellow River Basins, Northern China. In *Ancient Health: Skeletal Indicators of Agricultural and Economic Intensification*, ed. M. Cohen and G. Krane-Cramer. University Press of Florida: Orlando. 255–72.

Phadtare, N. R. 2000. 4000–3500 Cal Yr B.P. in the Central Higher Himalaya of India Based on Pollen Evidence from Alpine Peat. *Quaternary Research* 53:122–29.

Pietrusewsky, M. 1974. The Paleodemography of a Prehistoric Thai Population: Non Nok Tha. *Asian Perspectives* 17:125–40.

Pietrusewsky, M., and M. T. Douglas. 2001. Intensification of Agriculture at Ban Chiang: Is There Evidence from the Skeletons? *Asian Perspectives* 40:157–78.

Pietrusewsky, M., M. T. Douglas, and R. M. Ikehara-Quebral. 1997. An Assessment of Health and Disease in the Prehistoric Inhabitants of the Mariana Islands. *American Journal of Physical Anthropology* 104:315–42.

Pietrusewsky, M., and C. H. Tsang. 2003. A Preliminary Assessment of Health and Disease in Human Skeletal Remains from Shi San Hang: A Prehistoric Aboriginal Site on Taiwan. *Anthropological Science* 111:203–23.

Pilbeam, D., and S. J. Gould. 1974. Size and Scaling in Human Evolution. *Science* 186:892–901.

Pilcher, J. R., and V. A. Hall. 1992. Towards a Tephrochronology for the Holocene of the North of Ireland. *Holocene* 2:255–59.

Possehl, G. 2002. *The Indus Civilization: A Contemporary Perspective*. AltaMira Press: Lanham, Md.

Possehl, G. L., and P. C. Rissman. 1992. The Chronology of Prehistoric India from Earliest Times to the Iron Age. In *Chronologies in Old World Archaeology*, ed. R. W. Ehrich. University of Chicago: Chicago. 465–90.

Powell, M. L. 1988. *Status and Health in Prehistory: A Case Study of the Moundville Chiefdom*. Smithsonian Institution Press: Washington, D.C.

Prasad, S., and Y. Enzel. 2006. Holocene Paleoclimates of India. *Quaternary Research* 66:442–53.

Puri, G. S., R. K. Gupta, and V. M. P. S. Meher-Homji. 1989. *Forest Ecology*. Oxford & IBH Publishing: New Delhi.

Raczek, T. P. 2003. Subsistence Strategies and Burial Rituals: Social Practices in the Late Deccan Chalcolithic. *Asian Perspectives* 42:247–66.

Raikes, R. L., and G. F. Dales. 1986. Reposte to Wasson's Sedimentological Basis of the Mohenjo-Daro Flood Hypothesis. *Man and Environment* 10:33–44.

Rajaguru, S. N. 1988. Environment. In *Excavations at Inamgaon*, ed. M. K. Dhavalikar, H. D. Sankalia, and Z. D. Ansari. Deccan College Postgraduate Research Institute: Pune. 9–15.

Rajaguru, S. N., and V. S. Kale. 1985. Changes in Fluvial Regime of Western Maharashtra Upland Rivers During Late Quaternary. *Journal of Geological Social Indicators* 26:16–27.

Rathbun, T. A. 1987. Health and Disease at a South Carolina Plantation—1840–1870. *American Journal of Physical Anthropology* 74:239–53.

Ravizza, G., and B. Peucker-Ehrenbrink. 2003. Chemostratigraphic Evidence of Deccan Volcanism from the Marine Osmium Isotope Record. *Science* 302:1392–95.

Rewekant, A. 2001. Do Environmental Disturbances of an Individual's Growth and Development Influence the Later Bone Involution Processes? A Study of Two Mediaeval Populations. *International Journal of Osteoarchaeology* 11:433–43.

Rewekant, A., and B. Jerszynska. 1995. Patterns of Cortical Bone Growth in Children: An Example from Medieval Populations. *Anthropologie* 33:79–82.

Ribot, I., and C. Roberts. 1996. A Study of Non-Specific Stress Indicators and Skeletal Growth in Two Mediaeval Subadult Populations. *Journal of Archaeological Science* 23:67–79.

Robbins, G. 2007. Population Dynamics, Growth and Development in Chalcolithic Sites of the Deccan Plateau, India. Ph.D. dissertation, University of Oregon, Eugene.

———. In press. Don't Throw the Baby Out with the Bathwater: Estimating Fertility from Subadult Skeletons. *International Journal of Osteoarchaeology*.

Robbins, G., and L. Cowgill. 2009. Bayesian Approaches to Measuring Body Mass in Subadults from Kulubnarti, Grasshopper Pueblo, and Inamgaon. *American Journal of Physical Anthropology* 138:222.

Robbins, G., V. Mushrif, V. N. Misra, R. K. Mohanty, and V. S. Shinde. 2007. Report on the Human Remains at Balathal. *Man and Environment* 31:50–65.

Robbins, G., P. W. Sciulli, and S. Blatt. 2010. Estimating Body Mass in Subadult Human Skeletons. *American Journal of Physical Anthropology* 143:146–50.

Roberts, N., and H. E. Wright. 1993. Vegetational, Lake Level, and Climate History of the near East and Southwest Asia. In *Global Climates since the Last Glacial Maximum*, ed. H. E. Wright Jr., J. E. Kutzbech, T. I. Webb, W. F. Ruddiman, and F. A. Street-Perrot. University of Minnesota Press: Minneapolis.

Roy, P. D., R. Sinha, W. Smykatz-Kloss, A. K. Singhvi, and Y. C. Nagar. 2008. Playas of the Thar Desert: Mineralogical and Geochemical Archives of Late Holocene Climate. *Asian Journal of Earth Sciences* 1:43–61.

Ruff, C. B. 1994. Morphological Adaptation to Climate in Modern and Fossil Hominids. *Yearbook of Physical Anthropology* 37:65–107.

———. 1995. Biomechanics of the Hip and Birth in Early Homo. *American Journal of Physical Anthropology* 98:527–74.

———. 1998. Evolution of the Hominid Hip. In *Primate Locomotion: Recent Advances*, ed. E. Strasser, J. Fleagle, H. McHenry, and A. Rosenberger. Plenum Press: New York. 449–69.

———. 2000. Body Size, Body Shape, and Long Bone Strength in Modern Humans. *Journal of Human Evolution* 38:269–90.

———. 2002a. Long Bone Articular and Diaphyseal Structure in Old World Monkeys and Apes. I: Locomotor Effects. *American Journal of Physical Anthropology* 119:305–42.

———. 2002b. Variation in Human Body Size and Shape. *Annual Review of Anthropology* 31:211–32.

———. 2003a. Growth in Bone Strength, Body Size, and Muscle Size in a Juvenile Longitudinal Sample. *Bone* 33:317–29.

———. 2003b. Ontogenetic Adaptation to Bipedalism: Age Changes in Femoral to Humeral Length and Strength Proportions in Humans, with a Comparison to Baboons. *Journal of Human Evolution* 45:317–49.

———. 2005a. Growth Tracking of Femoral and Humeral Strength from Infancy through Late Adolescence. *Acta Paediatrica* 94:1030–37.

———. 2005b. Mechanical Determinants of Bone Form: Insights from Skeletal Remains. *Journal of Musculoskeletal Neuronal Interaction* 5:202–12.

———. 2007. Body Size Prediction from Juvenile Skeletal Remains. *American Journal of Physical Anthropology* 133:698–716.

Ruff, C. B., B. Holt, and E. Trinkaus. 2006. Who's Afraid of the Big Bad Wolff? "Wolff Is Law" and Bone Functional Adaptation. *American Journal of Physical Anthropology* 129:484–98.

Ruff, C. B., and J. A. Runestad. 1992. Primate Limb Bone Structural Adaptations. *Annual Review of Anthropology* 21:407–33.

Ruff, C. B., E. Trinkaus, A. Walker, and C. S. Larsen. 1993. Postcranial Robusticity in Homo I: Temporal Trends and Biomechanical Interpretation. *American Journal of Physical Anthropology* 91:21–53.

Ruff, C. B., and A. Walker. 1993. Body Size and Body Shape. In *The Nariokotome Homo Erectus Skeleton*, ed. A. Walker et al. Harvard University Press: Cambridge.

Ruff, C. B., A. Walker, and E. Trinkaus. 1994. Postcranial Robusticity in Homo, III: Ontogeny. *American Journal of Physical Anthropology* 93:35–54.

Sali, S. A. 1986. *Daimabad 1976–79*. Archaeological Survey of India: New Delhi.

Sankalia, H., S. B. Deo, and Z. D. Ansari. 1960. *From Prehistory to Protohistory at Nevasa*. Deccan College Research Institute: Pune.

———. 1971. *Chalcolithic Navdatoli*. Deccan College: Poona.

Sattenspiel, L., and H. Harpending. 1983. Stable Populations and Skeletal Age. *American Antiquity* 48:489–98.

Saunders, S. R. 2000. Subadult Skeletons and Growth-Related Studies. In *Biological Anthropology of the Human Skeleton*, ed. M. A. Katzenberg and A. Saunders. Wiley-Liss: New York. 135–62.

Saunders, S. R., and L. Barrans. 1999. What Can Be Done About the Infant Category in Skeletal Samples? In *Human Growth in the Past: Studies of Bones and Teeth*, ed. R. B. Hoppa and C. M. Fitzgerald. Cambridge University Press: New York. 183–209.

Saunders, S. R., R. D. Hoppa, and R. Southern. 1993. Diaphyseal Growth in a Nineteenth Century Skeletal Sample of Subadults from St Thomas Church, Belleville, Ontario. *International Journal of Osteoarchaeology* 3:265–81.

Saunders, S. R., D. A. Herring, and G. Boyce. 1995. Can Skeletal Samples Accurately Represent the Living Population They Come From? The St. Thomas Cemetery Site, Belleville, Ontario. In *Bodies of Evidence: Reconstructing History through Skeletal Analysis*, ed. A. L. Grauer. Wiley-Liss: New York. 69–89.

Saunders, S. R., A. Herring, L. Sawchuk, G. Boyce, R. Hoppa, and S. Klepp. 2002. The St. Thomas Anglican Church Project. In *The Backbone of History: Health and Nutrition in the Western Hemisphere*, ed. R. H. Steckel and J. Rose. Cambridge University Press: New York.

Scheuer, J. L., and S. Black. 2000. *Developmental Juvenile Osteology*. Academic Press: New York.

Scheuer, J. L., J. H. Musgrave, and S. P. Evans. 1980. Estimation of Perinatal and Late Foetal Age from Limb Bone Length by Linear and Logarithmic Regression. *Annals of Human Biology* 7:257–65.

Schumm, S. A. 1993. River Response to Base Level Change: Implications for Sequence Stratigraphy. *Journal of Geology* 101:279–94.

Sharma, S., M. Joachimski, M. Sharma, H. J. Tobschall, I. B. Singh, C. Sharma, M. S. Chauhan, and G. Morgenroth. 2004. Late Glacial and Holocene Environmental Changes in Ganga Plain, Northern India. *Quaternary Science Reviews* 23:145–59.

Sharma, S., M. Joachimski, H. J. Tobschall, I. B. Singh, D. P. Tewari, and R. Tewari. 2004. Oxygen Isotopes of Bovid Teeth as Archives of Paleoclimatic Variations in Archaeological Deposits of the Ganga Plain, India. *Quaternary Research* 62:19–28.

Sherwood, R. J., R. S. Meindl, H. B. Robinson, and R. L. May. 2000. Fetal Age: Methods of Estimation and Effects of Pathology. *American Journal of Physical Anthropology* 113:305–15.

Shinde, V. S. 1984. Early Settlements in the Central Tapi Basin. Ph.D. dissertation, University of Poona, Pune.

———. 1985. Kaothe. *Bulletin of the Deccan College Postgraduate Research Institute* 44:173–77.

———. 1989. New Light on the Origin, Settlement System and Decline of the Jorwe Culture of the Deccan, India. *South Asian Studies* 5:60–72.

———. 1990. Settlement Pattern of the Savalda Culture: The First Farming Community of Maharashtra. *Bulletin of the Deccan College Post-Graduate Research institute* 49:417–26.

———. 2000. The Origin and Development of the Chalcolithic in Central India. *Bulletin of the Indo-Pacific Prehistory Association* 19:115–24.

———. 2002. The Emergence, Development and Spread of Agricultural Communities in South Asia. In *The Origins of Pottery and Agriculture*, ed. Y. Yasuda. Lustre Press and Roli: New Delhi. 89–115.

Singh, G. 1971. The Indus Valley Culture Seen in the Context of Postglacial Climatic and Ecological Studies in North-West India. *Archaeology and Physical Anthropology of Oceania* 6:177–89.

Singh, G., S. K. Chopra, and A. B. Singh. 1973. Pollen-Rain from the Vegetation of North-West India. *Phytologist* 72:191–206.

Singh, G., R. D. Joshi, S. K. Chopra, and A. B. Singh. 1974. Late Quaternary History of Vegetation and Climate of the Rajasthan Desert, India. *Philosophical Transactions of the Royal Society of London Series B, Biological Sciences (1934–1990)* 267:467–501.

Singh, G., R. D. Joshi, and A. B. Singh. 1972. Stratigraphic and Radiocarbon Evidence for the Age and Development of Three Salt Lake Deposits in Rajasthan, India. *Quaternary Research* 2:496–505.

Singh, G., R. J. Wasson, and D. P. Agrawal. 1990. Vegetational and Seasonal Climatic Changes since the Last Full Glacial in the Thar Desert, Northwestern India. *Review of Paleobotany and Palynology* 64:351–58.

Sinha, R., W. Smykatz-Kloss, D. Stüben, S. P. Harrison, Z. Berner, and U. Kramar. 2006. Late Quaternary Palaeoclimatic Reconstruction from the Lacustrine Sediments of the Sambhar Playa Core, Thar Desert Margin, India. *Palaeogeography, Palaeoclimatology, Palaeoecology* 233:252–70.

Skinner, M. 1996. Developmental Stress in Immature Hominines from Late Pleistocene Eurasia: Evidence from Enamel Hypoplasia. *Journal of Archaeological Science* 23:833–52.

Skinner, M., and J. T. W. Hung. 1989. Social and Biological Correlates of Localized Enamel Hypoplasia of the Human Deciduous Cancine Tooth. *American Journal of Physical Anthropology* 79:159–75.

Skinner, M., and E. Newell. 2000. A Re-Evaluation of Localized Hypoplasia of the Primary Canine as a Marker of Craniofacial Osteopenia in European Upper Paleolithic Infants. *Acta Universitas Carol (Praha)* 41:41–58.

Smith, E. A., and B. Winterhalder. 1992. *Evolutionary Ecology and Human Behavior*. Aldine de Gruyter: New York.

Smith, H. B. 1991. Dental Development and the Evolution of Life History in Hominidae. *American Journal of Physical Anthropology* 86:157–74.

Smith, P., and L. K. Horowitz. 2007. Ancestors and Inheritors: A Bioanthropological Perspective on the Transition to Agropastoralism in the Southern Levant. In *Ancient Health: Skeletal Indicators of Agricultural and Economic Intensification*, ed. M. Cohen and G. Crane-Kramer. University Press of Florida: Orlando.

Staubwasser, M., and H. Weiss. 2006. Holocene Climate and Cultural Evolution in Late Prehistoric–Early Historic West Asia. *Quaternary Research* 66:372–87.

Steckel, R. H. 1995. Stature and the Standard of Living. *Journal of Economic Literature* 33:1903–40.

Steckel, R. H., and J. C. Rose. 2002. *The Backbone of History: Health and Nutrition in the Western Hemisphere*. Cambridge University Press: New York.

Steckel, R. H., J. C. Rose, C. S. Larsen, and P. L. Walker. 2002. Skeletal Health in the Western Hemisphere from 4000 B.C. to the Present. *Evolutionary Anthropology* 4:142–55.

Stock, J. 2006. Hunter-Gatherer Postcranial Robusticity Relative to Patterns of Mobility, Climatic Adaptation, and Selection for Tissue Economy. *American Journal of Physical Anthropology* 131:194–204.

Stock, J., and S. Pfeiffer. 2001. Linking Structural Variability in Long Bone Diaphyses to Habitual Behaviors: Foragers from the Southern African Later Stone Age and the Andaman Islands. *American Journal of Physical Anthropology* 115:337–48.

———. 2004. Long Bone Robusticity and Subsistence Behaviour among Later Stone Age Foragers of the Forest and Fynbos Biomes of South Africa. *Journal of Archaeological Science* 31:999–1013.

Storey, R. 1992. *Life and Death in the Ancient City of Teotihuacan: A Modern Paleodemographic Synthesis*. University of Alabama Press: Tuscaloosa.

Suganan, V. V. 1995. *Reservoir Fisheries of India*. Rome: Food and Agriculture Organization of the United Nations. June 23, 2006. http://www.fao.org/docrep/003/V5930E/V5930E06.htm/.

Sumner, D. R. 1984. *Size, Shape and Bone Mineral Content in the Human Femur in Growth and Aging.* University of Arizona: Tucson.

Sumner, D. R., and T. P. Andriacchi. 1996. Adaptation to Differential Loading: Comparison of Growth-Related Changes in Cross-Sectional Properties of the Human Femur and Humerus. *Bone* 19:121–26.

Swain, A. M., J. E. Kutzbach, and S. Hastenrath. 1983. Monsoon Climate of Rajasthan for the Holocene: Estimates of Precipitation Based on Pollen and Lake Levels. *Quaternary Research* 19:1–17.

Tanner, J. M. 1986. Normal Growth and Techniques of Growth Assessment. *Clinical Endocrinological Metabolism* 15:411–51.

Tayles, N. 1992. The People of Khok Phanom Di: Health as Evidence of Adaptation in a Prehistoric Southeast Asian Population. Ph.D. dissertation, University of Otago, New Zealand.

———. 1996. Anemia, Genetic Diseases, and Malaria in Prehistoric Mainland Southeast Asia. *American Journal of Physical Anthropology* 101:11–27.

Tayles, N., K. Domett, and K. Nelsen. 2000. Agriculture and Dental Caries? The Case of Rice in Prehistoric Southeast Asia. *World Archaeology* 32:68–83.

Telkka, A., A. Palkama, and P. Virtama. 1962. Prediction of Stature from Radiographs of Long Bones in Infants and Children. *Journal of Forensic Science* 7:474–79.

Thomas, P. K. 1988. Faunal Assemblage. In *Excavations at Inamgaon*, ed. M. K. Dhavalikar, H. D. Sankalia, and Z. D. Ansari. Deccan College Post-Graduate Research Institute: Pune. 823–963.

Tocheri, M. W., T. L. Dupras, P. Sheldrick, and J. E. Molto. 2005. Roman Period Fetal Skeletons from the East Cemetery (Kellis 2) of Kellis, Egypt. *International Journal of Osteoarchaeology* 15:326–41.

Trinkaus, E. 1981. Neanderthal Limb Proportions and Cold Adaptation. In *Aspects of Human Evolution*, ed. C. B. Stringer. Taylor and Francis: London. 187–219.

Trinkaus, E., S. E. Churchill, and C. B. Ruff. 1994. Postcranial Robusticity in Homo: Ii. Humeral Bilateral Asymmetry and Bone Plasticity. *American Journal of Physical Anthropology* 93:1–34.

Van der Meulen, M., M. W. Ashford, B. J. Kiratli, L. K. Bachrach, and D. R. Carter. 1996. Determinants of Femoral Geometry and Structure During Adolescent Growth. *Journal of Orthopedic Research* 14:22–29.

Van der Meulen, M., G. S. Beaupré, and D. R. Carter. 1993. Mechanobiologic Influences in Long Bone Cross-Sectional Growth. *Bone* 14:635–42.

Van Gerven, D. P., J. R. Hummert, and D. B. Burr. 1985. Cortical Bone Maintenance and Geometry of the Tibia in Prehistoric Children from Nubia's Batn El Hajar. *American Journal of Physical Anthropology* 66:275–80.

Van Gerven, D. P., S. G. Sheridan, and W. Y. Adams. 1995. The Health and Nutrition of a Medieval Nubian Population: The Impact of Political and Economic Change. *American Anthropology* 97:468–80.

Venkataraman, R., and S. T. Chari. 1953. Food Value of the Edible Portion of the Indian Chank Xancus Pyrum. *Proceedings of the National Academy of Science (Indian Academy of Sciences)* 22:22–23.

Vidarbha Irrigation Development Corporation, Nagpur. 2009. *Brief Information of Maharashtra*, ed. Maharashtra Govt.

Vishnu-Mittre, A. S. 1981. Wild Plants in Indian Folk Life: A Historical Perspective. In *Glimpses of Indian Ethnobotany*, ed. S. K. Jain. Botanical Survey of India: New Delhi. 37–58.

Von Rad, U., M. Schaaf, K. H. Michels, H. Schulz, W. H. Berger, and F. Sirocko. 1999. A 5000-Yr Record of Climate Change in Varved Sediments from the Oxygen Minimum Zone Off Pakistan Northeastern Arabian Sea. *Quaternary Research* 51:39–53.

Waldron, T. 1987. The Relative Survival of the Human Skeleton: Implications for Palaeopathology. In *Death, Decay and Reconstruction: Approaches to Archaeology and Forensic Science*, ed. A. Boddington, A. N. Garland, and R. C. Janaway. Manchester University Press: Manchester. 55–64.

Walimbe, S. R. 1986. The Burials. In *Daimabad, 1976–1979*, ed. S. A. Sali. Archaeological Survey of India, Government of India: New Delhi. 166–205, 641–740.

Walimbe, S. R., and P. B. Gambhir. 1997. *Long Bone Growth in Infants and Children: Assessment of the Nutritional Status*. Mujumdar Publications: Mangalore, India.

Wasson, R. J., G. I. Smith, and D. P. Agarwal. 1984. Late Quaternary Sediments, Minerals, and Inferred Geochemical History of Didwana Lake. *Paleogeography, Paleoclimatology, Paleoecology* 46:345–72.

Weiss, K. M. 1973. Demographic Models for Anthropology. *Society for American Archaeology*: Washington, D.C.

Wescott, D. J. 2006. Effect of Mobility on Femur Midshaft External Shape and Robusticity. *American Journal of Physical Anthropology* 130:201–13.

Wescott, D. J., and D. L. Cunningham. 2006. Temporal Changes in Arikara Humeral and Femoral Cross-Sectional Geometry Associated with Horticultural Intensification. *Journal of Archaeological Science* 33:1022–36.

Wheeler, M. 1959. *Early India and Pakistan: To Ashoka*. Praeger: New York.

Whittington, S. L., and D. M. Reed. 1997. *Bones of the Maya: Studies of Ancient Skeletons*. Smithsonian Institution Press: Washington, D.C.

Wood, J. W., G. R. Milner, H. C. Harpending, and K. M. Weiss. 1992. The Osteological Paradox: Problems of Inferring Prehistoric Health from Skeletal Samples. *Current Anthropology* 33:343–70.

Wright, H. E. 1993. *Global Climates since the Last Glacial Maximum*. University of Minnesota Press: Minneapolis.

Wright, L., and C. Yoder. 2003. Recent Progress in Bioarchaeology: Approaches to the Osteological Paradox. *Journal of Archaeological Research* 11:1059–61.

Yasuda, Y., V. Shinde, S. K. N. B. Kenkyu, and Geo-Genom Project. 2004. *Monsoon and Civilization*. Lustre Press, Roli Books: New Delhi.

Index

age estimation, 69–75, 128–41, 144; and body mass formulas, 86, 101–2; children, 72–74; infants, 70, 72–74; long bone lengths, 70–72; methods, 24; perinates, 70–72, 73. *See also* Daimabad; Inamgaon; Nevasa

agriculture, 1, 3, 4, 118; and human health in general, 10–12, 14, 22–23, 91; and monsoon, 28; and sedentism, 55; at Daimabad, 46–47, 59 (*see also* Daimabad); at Inamgaon, 48–51, 55, 60, 100, 109–11, 122 (*see also* Inamgaon); at Nevasa, 42, 59 (*see also* Nevasa); Black Cotton Soil, 28; in India today, 27–28, 38, 144; in the Deccan Chalcolithic, 7–9, 12–17, 19–22, 36, 38, 96–97, 100–111, 114–17, 119, 122–23; populations, xiii, 15, 100, 109; subsistence, 10–11, 14–15, 100, 116

artifacts, 6, 41, 62, 50, 57, 143; beads, 43–44, 47–48, 57–58, 116, 127; bone, 56, 57, 58, 60; bronze, 4, 44; ceramics (and pottery), 4, 5, 10, 33, 38, 41, 44, 47, 50, 60, 116, 117, 127, 144–45; copper, 3–5, 38, 43–44, 50, 57–58, 60, 116–17, 127; decorated pot sherds, 41; groundstone, 116–17, 127; shell, 56–58; stone tools, 39, 41, 57, 60; terracotta figurines, 56, 60; tool production, raw materials, 23, 144

barley, 7, 14, 19, 20, 28, 36, 38, 50–51, 55, 59–60, 96, 116, 119. *See also* agriculture; floral remains; subsistence transition

biocultural stress, 10, 12, 13, 15–16, 17–18, 19, 20–24, 36, 60, 80–81, 82, 88, 89–91, 93–100, 109–11, 112, 119–23

biodemography model, 18–22, 23, 78–81, 82, 97, 114–23

Black Cotton Soil. *See* agriculture

body mass, 20, 21, 23, 24, 82–85, 88, 89, 90–94, 99, 101, 103, 121; formulas 86–88; body mass index, 21, 24, 86–88, 92–96, 100, 103, 107, 109–11, 121–22; population differences, 92–94, 103–10, 111–12

burial, 4, 5, 19, 23, 42–44, 47–48, 58, 63, 80, 116, 117, 125–27, 143–44. *See also* Daimabad; Early Jorwe; Inamgaon; Late Jorwe; Nevasa

caries, 11, 13–17

climate, 6, 23, 24, 25, 28–35, 56, 143; change, xv, 9–12, 18–19, 21, 24, 28–35, 53–54, 59, 96, 111, 114–19, 122; paleoclimate, xiv, 18–19, 21, 23–24, 29–35, 36–37, 38, 53, 96, 111, 114–19. *See also* climate-culture change model

climate-culture change model, 6–12, 19–22, 34–37, 48–53, 96, 99, 111, 114–19, 122–23. *See also* climate

cow (cattle, or *Bos indicus*), 8, 42, 46,
  48, 51–52, 55–56, 60, 116–17; dung, 41.
  See also Early Jorwe; faunal remains;
  subsistence transition

Daimabad, 4, 6, 7–8, 12, 15, 16, 17, 18,
  19, 20–21, 38, 89, 97, 143, 144, 145;
  abandonment, 59–60, 119–21, 122; age
  estimation, 69, 70, 71, 73, 74, 75; archaeology, 44–48; artifacts, 41, 44, 46;
  burials, 43, 47–48, 63, 125–27; evidence
  of climate, 32; date of occupation, 44;
  excavation season dates, 44; faunal
  remains, 42; floral remains, 46; long
  bone growth, 13–14, 100–110; map, 3,
  45; paleodemography, 74–80, 98–99,
  120–21; settlement size, 47, 59–60, 75,
  76, 78, 79, 80, 116; subsistence, 42, 46,
  48. *See also* agriculture; climate; Deccan Chalcolithic; demography; Early
  Jorwe; Late Jorwe; long bone length
Deccan Chalcolithic, 2–4, 6, 23, 28, 37,
  38, 44, 48, 53, 57, 63, 82, 89, 111, 112, 114,
  116–23; dates, 3, 5; map, 2, 3; origins,
  4–6; skeletal populations, 12–13, 43,
  62, 69, 72, 74, 76, 77, 79, 88, 91, 92,
  96, 97, 100, 102, 105–7, 110, 111; sites,
  6, 9–10, 41; subsistence, 57. *See also*
  biodemography model; body mass;
  climate; climate-culture change model;
  Daimabad; demography; Early Jorwe;
  Inamgaon; Late Jorwe; Nevasa; subsistence transition
demography, 12, 13, 17–18, 20–21, 39,
  61–64, 69, 97–98, 110, 122, 143, 145;
  fertility, xiii, 14, 17–18, 20–22, 37, 55,
  61–65, 66, 67, 69, 72–73, 75–78, 79–80,
  97–99, 110, 119–22, 143; mortality, xiii,
  12, 17–18, 20–22, 37, 60, 61–62, 64, 75,
  77, 79, 80, 97, 98–99, 100, 110, 119–22,
  143; paleodemography, 18, 23, 61–64,
  69, 70, 74, 78, 119, 143. *See also* age
  estimation; biodemography model;

Daimabad; Deccan Chalcolithic;
  Inamgaon; Nevasa
dental defects. *See* enamel defects; LHPC
dental development, 12, 15–16, 17, 70, 72,
  74, 98, 100–102, 106, 145
developmental stress. *See* biocultural
  stress

Early Jorwe, 18–21, 71, 110, 114–22; burial
  practice, 47–48, 58; climate, 32,
  36–37, 38, 54; Daimabad, 7, 44, 46–48,
  59–60; dates, 6, 48; demographic
  profile of, 74–80; emaciation rates in,
  100–110; houses, 54, 63; Inamgaon,
  7, 48, 50–53, 60; Nevasa, 39, 59–60;
  previous bioarchaeology research on,
  14–18, 97–100; sites, 6–7; subsistence,
  7–8, 12, 46–47, 49–53, 55–57. *See also*
  agriculture; artifacts; biodemography
  model; climate; climate-culture change
  model, Daimabad; Deccan Chalcolithic; Inamgaon; Late Jorwe; Nevasa;
  subsistence transition
emaciation, 22, 88, 91, 92, 96, 99–100,
  110–11, 121
enamel defects, 15–18, 90, 98–100, 110. *See
  also* LHPC

faunal remains (animal bones), 8,
  31; at Daimabad, 46; at Inamgaon,
  51–53, 56–57; at Nevasa, 42. *See also*
  Daimabad; Deccan Chalcolithic; Early
  Jorwe, subsistence; Late Jorwe, subsistence; Inamgaon; Nevasa; subsistence
  transition
fertility. *See* demography
floral remains (plants), 8, 33–34, 46, 50,
  143–44

goats, 8, 10, 42, 46, 48, 52, 56, 60, 116–17.
  See also cow; faunal remains; Late
  Jorwe; sheep; subsistence transition
Gross Reproductive Rate, 62–63, 64–69

house, 3, 5, 8, 19, 38, 47–48, 50, 54–55, 57, 63, 80, 116, 127

Inamgaon, 6–8, 10, 12, 19–20, 23, 38, 58–60, 82, 89, 96–97, 100, 114, 116–22, 143–45; age estimation, 69–75, 101–2, 129–30, 133, 135; archaeology, 19–20, 48–53; artifacts, 41; burials, 12, 41, 47, 63; dental anthropology, 14–18, 21–22, 97–99, 110–11, 122; evidence of climate, 32; date of occupation, 48; excavation season dates, 48; faunal remains, 51–53; floral remains, 50–51; houses, 50; long bone growth, 13–14, 100–110; map, 3, 49; pathology, 13; new interpretations of archaeological record, 53–59; paleodemography, 20–22, 63, 74–78, 80–81, 121–23; settlement size, 20, 60, 75, 77–78, 80, 116; subsistence, 50–53. *See also* agriculture; biodemography model; climate; climate-culture change model; Deccan Chalcolithic; demography; Early Jorwe; Late Jorwe; long bone length; subsistence transition; subsistence transition
Indus civilization, 1–3, 4–5, 9, 34, 36, 38, 44, 46, 63, 118, 144

Jorwe, phase, 3, 4, 6–10; site, 6. *See also* Early Jorwe; Late Jorwe

Late Jorwe, 7–10, 48, 50–60, 71, 96, 114, 143; burial practice, 58–59, 117; climate, 8–10, 36, 37, 55, 59–60, 118; dates, 6, 48, 144; demographic profile of, 18, 20–21, 63, 75–76, 78–80, 119, 121; emaciation rates in, 21, 24, 82, 88, 99–100, 102–12, 121–22; houses, 50, 54; previous bioarchaeology research on, 11–16, 97, 99, 122; settlement size, 54, 60, 80; sites, 6–7; subsistence, 7–10, 16–17, 19, 38–39, 50–57, 59–60, 116–17, 119, 143. *See also* agriculture; artifacts; biodemography

model; climate; climate-culture change model, Daimabad; Deccan Chalcolithic; Inamgaon; Late Jorwe; Nevasa; subsistence transition
LEH, 15–17, 96, 122. *See also* enamel defects
LHPC, 15–17, 80–81, 97–99, 110, 122–23. *See also* enamel defects
life expectancy at birth. *See* mortality
long bone length, 20–21, 70–72, 74, 82, 85, 88, 91–92, 100–110, 119–20, 145; previous analysis, 13–14. *See also* age estimation

monsoon, 26, 28–33, 35–36, 38, 51, 56, 59, 116. *See also* agriculture; climate
mortality. *See* demography

Nevasa, 6, 12, 20, 23, 38, 82, 89; abandonment, 59–60, 119–22; age estimation, 69–74, 138–41; archaeology, 19, 39; artifacts, 41–42; burials, 12, 42–44, 63, 144; dental anthropology, 15–18, 97–100, 122; evidence of climate, 32; date of occupation, 39, 143; excavation season dates, 39; faunal remains, 42; floral remains, 42; long bone growth, 13–14, 100–110, 145; map, 3, 40; paleodemography, 20–21, 24, 74–76, 79, 120; relationship to Daimabad, 7, 20, 42, 44, 47, 59–60, 79, 119–20; settlement size, 47, 59–60, 75, 76, 78, 79, 80, 116; subsistence, 42. *See also* agriculture; climate; Deccan Chalcolithic; demography; Early Jorwe; Late Jorwe; long bone length

osteological paradox, 17–18, 22, 97–101, 122

paleoclimate. *See* climate
paleodemography. *See* demography

pathology, 13, 17, 82, 100
percent Cortical Area (%CA), 94–95

settlement, 1, 3–5, 6, 7–8, 23, 25, 32, 36, 112, 143, 144; abandonment, 8–12, 19–21, 34, 54–55, 59–60, 78–80, 114, 116, 118–19, 120–23; demography, 61–62, 69–80, 120; growth, 3, 7, 20–22, 47, 60, 76–80, 116, 120
sheep, 8, 10, 42, 46, 48, 52–53, 56, 60, 116–17. *See also* cow; faunal remains; goats; Late Jorwe; subsistence transition
stress. *See* biocultural stress
subsistence transition, 6–12, 14–17, 19–22, 50–51, 55, 59, 80–81, 89, 96–97, 99–100, 103, 109–13. *See also* agriculture; Daimabad; Early Jorwe; floral remains; faunal remains; Inamgaon; Late Jorwe; Nevasa

Gwen Robbins Schug is associate professor of biological anthropology at Appalachian State University. She is author of several recent publications on skeletal evidence for leprosy in India during the second millennium B.C., human versus nonhuman bone identifications for the Donner party hearth assemblage, and methodological contributions on body mass estimation in subadult human skeletons and a technique for fertility-centered paleodemography in skeletal assemblages. Her research has been supported by two Fulbright fellowships, the American Institute of Indian Studies, and the George Franklin Dales Foundation. In 2010, she received the William C. Strickland Outstanding Young Faculty Award at Appalachian State University.

Bioarchaeological Interpretations of the Human Past: Local, Regional, and Global Perspectives
Edited by Clark Spencer Larsen

This series examines the field of bioarchaeology, the study of human biological remains from archaeological settings. Focusing on the intersection between biology and behavior in the past, each volume will highlight important issues, such as biocultural perspectives on health, lifestyle and behavioral adaptation, biomechanical responses to key adaptive shifts in human history, dietary reconstruction and foodways, biodistance and population history, warfare and conflict, demography, social inequality, and environmental impacts on population.

Ancient Health: Skeletal Indicators of Agricultural and Economic Intensification, edited by Mark Nathan Cohen and Gillian M. M. Crane-Kramer (2007; first paperback edition, 2012)

Bioarchaeology and Identity in the Americas, edited by Kelly J. Knudson and Christopher M. Stojanowski (2009; first paperback edition, 2010)

Island Shores, Distant Pasts: Archaeological and Biological Approaches to the Pre-Columbian Settlement of the Caribbean, edited by Scott M. Fitzpatrick and Ann H. Ross (2010)

The Bioarchaeology of the Human Head: Decapitation, Decoration, and Deformation, edited by Michelle Bonogofsky (2011; first paperback edition, 2015)

Bioarchaeology and Climate Change: A View from South Asian Prehistory, by Gwen Robbins Schug (2011; first paperback edition, 2017)

Violence, Ritual, and the Wari Empire: A Social Bioarchaeology of Imperialism in the Ancient Andes, by Tiffiny A. Tung (2012; first paperback edition, 2013)

The Bioarchaeology of Individuals, edited by Ann L. W. Stodder and Ann M. Palkovich (2012; first paperback edition 2014)

The Bioarchaeology of Violence, edited by Debra L. Martin, Ryan P. Harrod, and Ventura R. Pérez (2012; first paperback edition 2013)

Bioarchaeology and Behavior: The People of the Ancient Near East, edited by Megan A. Perry (2012)

Paleopathology at the Origins of Agriculture, edited by Mark Nathan Cohen and George J. Armelagos (2013)

Bioarchaeology of East Asia: Movement, Contact, Health, edited by Kate Pechenkina and Marc Oxenham (2013)

Mission Cemeteries, Mission Peoples: Historical and Evolutionary Dimensions of Intracemetery Bioarchaeology in Spanish Florida, by Christopher M. Stojanowski (2013)

Tracing Childhood: Bioarchaeological Investigations of Early Lives in Antiquity, edited by Jennifer L. Thompson, Marta P. Alfonso-Durruty, and John J. Crandall (2014)

The Bioarchaeology of Classical Kamarina: Life and Death in Greek Sicily, by Carrie L. Sulosky Weaver (2015)

Victims of Ireland's Great Famine: The Bioarchaeology of Mass Burials at Kilkenny Union Workhouse, by Jonny Geber (2016)

Colonized Bodies, Worlds Transformed: Towards a Global Bioarchaeology of Contact and Colonialism, edited by Melissa S. Murphy and Haagen D. Klaus (2017)

Bones of Complexity: Bioarchaeological Case Studies of Social Organization and Skeletal Biology, edited by Haagen D. Klaus, Amanda R. Harvey, and Mark N. Cohen (2017)

www.ingramcontent.com/pod-product-compliance
Lightning Source LLC
Chambersburg PA
CBHW031431270326
41930CB00007B/662